IET

Code of Practice
for Low and Extra Low Voltage Direct Current Power Distribution in Buildings

Published by: The Institution of Engineering and Technology, London, United Kingdom

The Institution of Engineering and Technology is registered as a Charity in England & Wales (no. 211014) and Scotland (no. SC038698).

First published 2015

The Institution of Engineering and Technology
Michael Faraday House
Six Hills Way, Stevenage
Herts, SG1 2AY, United Kingdom

www.theiet.org

ISBN 978-1-84919-835-6 (paperback)
ISBN 978-1-84919-836-3 (electronic)

Typeset in the UK by Phoenix Photosetting, Chatham, Kent
Printed digitally by Elanders UK

Contents

Introduction

Low voltage direct current power systems exist today in our wider national infrastructure and industry. The development of IEEE standards for remote powering of devices over structured cabling infrastructures presents a unique opportunity to consider a holistic approach to a wider provision of d.c power within the built environment. Adopting such an approach will provide a reduction in the usage of building resources and create a step change in our thinking about the distribution and conversion of power.

NOTE: For the purposes of this Code of Practice, the term low voltage, unless otherwise stated, includes both extra-low voltage (ELV) and low voltage (LV) as defined in BS 7671. Where a specific voltage band is described, the terms LV or ELV are used as appropriate (see Section 2).

This trend is, in part, based on the use of telecommunications cabling infrastructure, which was not initially designed or installed for the delivery of power. As a result, there is a need for recommendations that would cover both the control of power over legacy cabling and the design of future installations. This is particularly important as much of the growth in this area is based upon the attachment of remote IP-enabled devices, providing separated extra-low voltage (SELV) circuits, which are compliant with IEEE 802.3 specifications. Such systems are generically termed Power over Ethernet (PoE), which are evolving to increase the power level delivered over each cable. This Code of Practice addresses this in Section 4.

However, not all the implementations of ELV d.c. power distribution over telecommunications cabling infrastructures adopt the IEEE solutions. Some of these implementations are clearly proprietary but use power levels lower than those of the IEEE specifications, while others claim compliance but deliver power levels significantly in excess of those specifications. As a result, greater awareness needs to be applied in such circumstances, in relation both to the cabling and to the devices supplying and receiving the power. This Code of Practice addresses this in Section 5.

Substantial investigations have been undertaken into the heating effects of power delivery over telecommunications cabling infrastructures in accordance with the recognised structure (or generic) cabling standards in the BS EN 50173, ISO/IEC 11801 and ANSI/TIA-568 series. However, not all cables used in telecommunications cabling infrastructures are compliant with these standards and these may experience significantly higher than expected temperature rises for a given level of current. These are also considered in Sections 4 and 5.

Some powering solutions use cabling infrastructures that are of proprietary design (i.e. not standards-based) but that are installed explicitly for the delivery of power using d.c. These implementations may be within the ELV or LV bands. An example of the latter is the provision of 380–400 V d.c. supplies to equipment within data centres. Such proprietary solutions are addressed in Section 6 of this Code of Practice.

The growth in low voltage direct current powering solutions is based upon a combination of potential benefits for integrated management of services, reductions in energy consumption and improvements in energy efficiency. These benefits result from increasing granularity in the monitoring and control of space and energy usage and the use of equipment with greater operational energy efficiency. Certain solutions focus on the latter by converting existing mains power supply cabling to d.c. distribution and are addressed in Sections 7 and 8 of this Code of Practice.

This Code of Practice will be of interest to a wide range of stakeholders, including customers, owners and insurers of low voltage d.c. power applications.

Participants in the Technical Committee

The IET wishes to acknowledge the support received from representatives of the following organisations in the development of this Code of Practice.

Committee membership:
British Electrotechnical and Allied Manufacturers' Association (BEAMA)
Chartered Institution of Building Service Engineers (CIBSE)
Communications Management Association (part of BCS, the Chartered Institute of IT)
Custom Electronic Design & Installation Association (CEDIA)
Department of Energy & Climate Change (DECC)
Electrical Contractors' Association (ECA)
Electrical Contractors' Association of Scotland (SELECT)
Energy Services and Technology Association (ESTA)
Industry experts:
      Arup
      CommScope
      FES Ltd
      J Brand Ltd
      Moixa Technology Ltd
      Rolls-Royce plc
      TRW Conekt
Institution of Engineering and Technology (IET)
Knowledge Transfer Network Ltd (KTN)
National Association of Professional Inspectors and Testers (NAPIT)
Research institutions:
      Aston University
      University of Bath
      University of Liverpool
      University of Strathclyde
The City and Guilds of London Institute (City & Guilds)

Corresponding members:
Department for Communities & Local Government (DCLG)
Energy Networks Association (ENA)
ERA Technology
Health and Safety Executive (HSE)
Interserve Ltd
Lighting Industry Association (LIA)
NG Bailey Ltd
The Association for Automation, Instrumentation and Control and Laboratory Technology (GAMBICA)
Western Power Distribution (WPD)

Acknowledgements

The IET would like to thank the following parties for their contributions to this document:

Lead technical author:
Mike Gilmore FIET (e-Ready Building)

Additional contributors:

Bill Allan BEng(Hons) CEng MIET (NAPIT)

Frank Bertie MIET (NAPIT)

Ken Bromley (DCLG)

Bob Brotherton BEng(Hons) CEng MIET MCIBSE MIHEEM (FES Ltd)

Bob Cairney IEng MIET (SELECT)

James Campbell PhD (Rolls-Royce plc)

Mark Coates BEng (ERA Technology)

Mark Coles BEng(Hons) MIET (IET)

Michael Collinge MIET (NAPIT)

John Counsell PhD MIET (University of Liverpool)

Mark Dale (WPD)

Mike Daly BSc(Hons) (e-Ready Building)

Abdullah Emhemed PhD (University of Strathclyde)

Andrew Evans BEng(Hons) CEng MInstMC MIET (GAMBICA)

Roger Hazelden CPhys MInstP (TRW Conekt)

Graeme Hodgson EngD CEng MIET (Moixa Technology Ltd)

Paul Huggett PhD (KTN)

Blane Judd BEng FCGI CEng FIET FCIBSE FCIPHE (BLTK Consulting) *Chair*

Zhengyu Lin PhD CEng MIET MIEEE (Aston University)

Gary Middlehurst CEng MSc MCIBSE MIET MASHRAE (Advanced Intelligent Buildings Ltd)

Lindsay Moody BEng(Hons) MSc CEng MIET MCIBSE (NG Bailey Ltd)

K J Morton BSc CEng FIET (HSE)

Malcolm Mullins BA CEng FIET (Honeywell)

Tim Oldershaw MIET (J Brand Ltd)

Bernard Pratley (LIA)

Miles Redfern PhD BSc DMS CEng MIET (University of Bath)

Simon Robinson CEng FIET FCIBSE MIMechE MSLL (WSP Parsons Brinckerhoff)

Guy Singleton MIET LCGI MInstLM (ImagineThis)

Peter Smith (Interserve Ltd)

David Spillett (ENA)

Cameron Steel CEng FIET MCIBSE MInstRE (BK Design Associates)

David Stefanowicz CEng BEng(Hons) MInstMC MIET MIEEE (ECA)

Dani Strickland PhD MIET (Aston University)

Kris Szajdzicki BSc FEI (ND Metering Solutions)

Tuan C Tan PhD CEng FIET (CommScope)

Raj Vagdia MIET (BEAMA)

Salim Visram (City & Guilds)

Jon Warren MEng MBA CEng MIET (DECC)

Tom Waterson MEng(Hons) CEng MIET (Arup)

Bill Wright MA CEng FIET (ECA)

Chris Wright MA(Cantab) DiplArch (Moixa Technology Ltd)

Scope, document structure and fundamentals

1.1 Scope

The distribution of low voltage direct current within buildings may take a number of forms, including the use of:

(a)   telecommunications cabling in accordance with recognised standards to interconnect:
   (i)   power sources and powered devices (forming a SELV circuit) in accordance with recognised standards;
   (ii)   power sources and powered device interfaces of proprietary design;
             NOTE: for the purposes of this document, a proprietary design is one for which the specification is provided by a private individual or corporation (possibly under a trademark or patent)
(b)   wiring installed specifically for the purpose of low voltage d.c. power distribution connecting power sources and powered device interfaces;
(c)   existing low voltage alternating current (LV a.c.) wiring converted for the purpose of d.c. power distribution connecting power sources and powered device interfaces.

This Code of Practice sets out the requirements for the design, specification, selection, installation, commissioning, operation and maintenance of solutions in accordance with (a)(ii), (b) and (c). In addition, this Code of Practice suggests amendments to the reference installation designs and the operational management of telecommunications cabling in relation to power loads to support solutions in accordance with (a).

The power sources may be either supplied from LV a.c. or d.c. power supplies. The LV d.c. power supplies may also include other power components such as storage sub-systems.

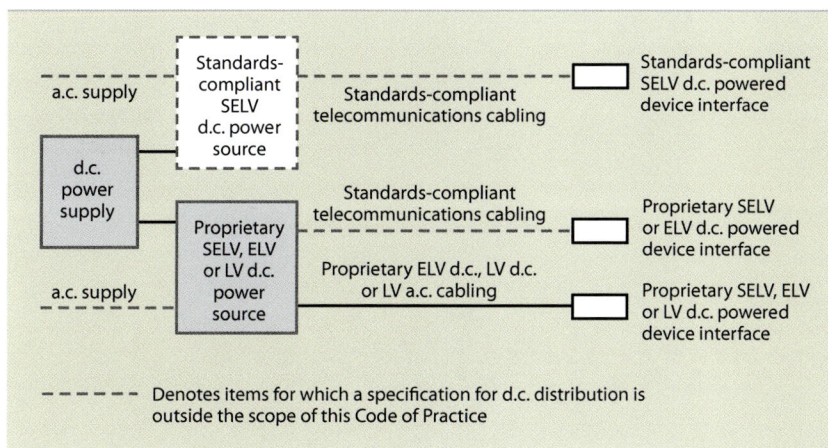

Figure 1: Schematic showing items within the scope of this Code of Practice

NOTE: All equipment (including power sources, wiring systems, protection and control devices) incorporated in work covered by the scope of this document – whether newly provided or adapted from previous use – is to be selected, erected, inspected and tested in accordance with the requirements of BS 7671.

© The Institution of Engineering and Technology

For the items within the scope as outlined above and detailed in Figure 1, this Code of Practice covers LV d.c.:

(d)   power principles and relevant standards;

(e)   power distribution over new and retrofit wiring/cabling (including switchgear and accessories);

(f)   protection for safety and equipment.

This Code of Practice does not address the design, installation or operation of LV d.c. power distribution in hazardous environments.

NOTE 1: The provision of power generated and supplied to the grid is not within the scope of this Code of Practice and relevant information is provided by external reference only.

NOTE 2: This Code of Practice does not provide in depth coverage of high voltage (HV) d.c., mobile or temporary d.c. installations or telecommunications systems, however, material may be relevant and applicable to interested parties.

1.2   Document structure

Section 3 introduces the justification of LV d.c. power distribution within buildings.

Section 4 addresses the use of telecommunication cabling with standards-based SELV power delivery solutions.

Section 5 addresses the use of telecommunication cabling with proprietary SELV and ELV power delivery solutions.

Section 6 addresses the use of proprietary cabling with proprietary SELV, ELV and LV d.c. power delivery solutions.

Section 7 addresses the use of conventional single-phase a.c. power distribution cabling with proprietary ELV and LV d.c. power delivery solutions.

Section 8 addresses the use of conventional 3-phase a.c. power distribution cabling with proprietary ELV and LV d.c. power delivery solutions.

1.3   Fundamentals

1.3.1   Regulations

All equipment (including power sources, wiring systems, protection and control devices) incorporated in work covered by the scope of this Code of Practice – whether newly provided or adapted from previous use – is to be selected, erected, inspected and tested in accordance with the requirements of BS 7671.

1.3.2   Awareness, qualifications and competence

Before engaging in the implementation of d.c powering solutions in accordance with this Code of Practice, personnel should be aware of the key differences between the handling of d.c and a.c circuits (the IET is preparing a Technical Briefing to address this topic).

With particular reference to implementations of Section 7 and Section 8 of this Code of Practice, personnel should:

(a)   have obtained any relevant qualifications appropriate to the type of installation to be undertaken (City & Guilds are developing specific d.c. related qualifications);

(b)   be deemed competent to undertake the installation (measures of competency are in development).

1.3.3   d.c. power sources

Some of the implementations of d.c. power distribution described in this Code of Practice may be supported by sources of d.c. power. These sources are assumed to have appropriate power quality characteristics and the necessary protection (for example, overload, short circuit, electric shock and thermal) for connection to the components of the d.c power distribution systems.

It is assumed that accommodation of d.c. sources is in accordance with the existing standards or Codes of Practice. Examples of such documents include:

(a)   BS EN 50272 series: *Safety requirements for secondary batteries and battery installations*;

(b)   IET Code of Practice: *Grid connected solar photovoltaic PV systems*.

1.3.4   Voltage levels

The reader's attention is drawn to the voltage ranges that are addressed in this Code of Practice. This is particularly relevant in Section 6, Section 7 and Section 8 and where voltages may exceed those typically encountered. The following icons are inserted at the beginning of each Section as a reminder to readers.

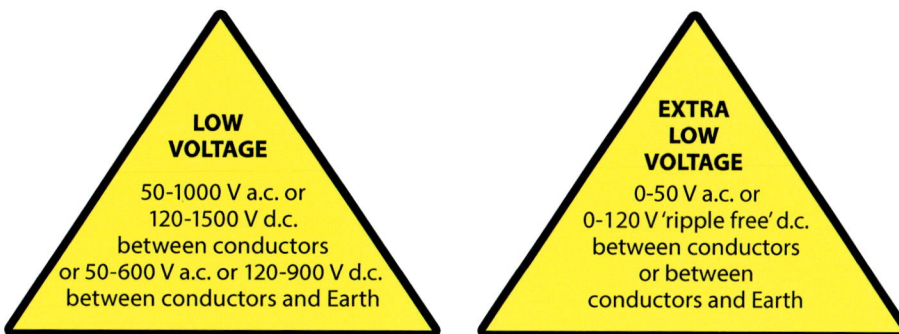

LOW
VOLTAGE
50-1000 V a.c. or
120-1500 V d.c.
between conductors
or 50-600 V a.c. or 120-900 V d.c.
between conductors and Earth

EXTRA
LOW
VOLTAGE
0-50 V a.c. or
0-120 V 'ripple free' d.c.
between conductors
or between
conductors and Earth

1.3.5   Galvanic corrosion

When surfaces of metals with different electro-chemical properties are connected together there will be a galvanic potential between these surfaces. The rate of corrosion depends on the:

(a)   electrochemical potential between the two metals;

(b)   conditions under which contact is made;

(c)   humidity and other environmental parameters in the vicinity of the connection.

© The Institution of Engineering and Technology

BS EN 60950-1 contains a table that states the electro-chemical potentials for a variety of metals. Electrochemical potentials of 300 mV (maximum) maintain a low galvanic effect in a moderately corrosive environment. However, lower potential differences are recommended in order to ensure a low impedance contact and a reliable electrical contact.

Further information is provided in BS EN 50310.

SECTION 2

Definitions and abbreviations

2.1 Definitions

Access network service provider

The operator of any service that furnishes telecommunications content (transmissions) delivered over the access network.

Channel

Between application-specific equipment and an external network interface, any transmission path comprising passive cabling components between application-specific equipment or between application-specific equipment and an external network interface.
[SOURCE: BS EN 50173-1:2011 modified]

Cord

Cable unit or element with a minimum of one termination.
[SOURCE: BS EN 50173-1:2011]

d.c. power distribution system

The power source, the powered device and the interconnecting equipment and cabling.

Extra-low voltage (ELV)

Not exceeding 50 V a.c. or 120 V ripple-free d.c., whether between conductors or to Earth.
[SOURCE: BS 7671]

Link

The transmission path between two specified test interfaces of generic cabling.
[SOURCE: BS 6701:2010]

Low voltage (LV)

Exceeding extra-low voltage but not exceeding 1,000 V a.c. or 1,500 V d.c. between conductors, or 600 V a.c. or 900 V d.c. between conductors and Earth.
[SOURCE: BS 7671]

Information technology

see telecommunications.
[SOURCE: BS EN 50173-1 and ISO/IEC 11801]

Proprietary

Implementation for which the specification is provided by a private individual or corporation (possibly under a trademark or patent).

Separated extra-low voltage (SELV) circuit

An extra-low voltage system that is electrically separated from Earth and from other systems in such a way that a single fault cannot give rise to the risk of electric shock.
[SOURCE: BS 7671]

Standards-based

Implementation for which the specification is contained within documents produced by national, regional or international standard-setting organisations.

Telecommunications

The branch of technology concerned with the transmission, emission and reception of signs, signals, writing, images and sounds; that is, information of any nature by cable, radio, optical or other electromagnetic systems.

NOTE: The term 'telecommunications' has no legal meaning when used in this Code of Practice.

[SOURCE: BS EN 50173-1:2011 and ISO/IEC 11801: Ed2.2]

Telecommunications cabling

The system of cables, cords and connecting hardware intended to support the operation of information technology equipment as, or as part of, a telecommunications system.
[SOURCE: BS 6701:2010]

Telecommunications system

A combination of telecommunications equipment and telecommunications cabling, which provides the distribution of telecommunications applications within and/or between premises.

NOTE 1: Examples of telecommunications applications are those which perform actions such as processing, displaying or transferring information including numerical data, text, audio, still or moving images together with any combination of these. Telecommunications applications are also used to control machinery, building facilities or other non-telecommunications equipment.

NOTE 2: In addition to telecommunications cabling, telecommunications systems also use waveguides (including optical waveguides) and radio transmission.

[SOURCE: BS 6701:2010]

2.2 Abbreviations

a.c.	alternating current
ADSL	Asymmetrical Digital Subscribers Line
BS	British Standard
CLC	CEN Electrotechnical Committee (European standardisation organisation)
d.c.	direct current
ELV	extra-low voltage
EN	European Norm
HVAC	Heating ventilation and air conditioning
IEEE	Institute of Electrical and Electronic Engineers
IP	Internet Protocol
ISDN	Integrated Services Digital Network
ISO/IEC	Joint committee of the International Standards Organisation and the International Electrotechnical Commission
IT	information technology
LV	low voltage
PD	powered device
PoE	power over Ethernet (in accordance with IEEE 802.3 standards)
PSE	power sourcing equipment
SELV	separated extra-low voltage

V a.c.   volt (alternating current)
V d.c.   volt (direct current)

2.3   Symbols

i_c d.c. current per conductor
N number of cables in a group or bundle
R resistance of a conductor (or the average resistance of the conductors in a circuit)

d.c. power distribution

3.1 The justification

Within buildings, the growth in the use of electronic devices, which distribute power internally via d. c. circuits, has encouraged the implementation of LV d.c. distribution to those devices in order to improve overall energy efficiency. The types and purposes of these implementations are growing rapidly due to five primary factors:

(a) the specification by standard-setting bodies (such as the IEEE) of remote powering solutions designed to operate over premises telecommunications cabling that has been designed and installed in accordance with recognised European, international and North American standards.

(b) although the IEEE specifications are predicated by the presence of Internet Protocol (IP) devices at the end of the d.c. powered distribution network, the opportunities offered by remote powering are also being used by others where the end-devices are not IP-based but the central management of those devices is. These IP-related developments have also contributed to a major advance in building services/management integration.

(c) the recognition that d.c. operation delivers more efficient and controllable outputs for end-devices that can offer energy efficiency advantages, including a range of solutions such as:
  (i) ELV d.c. distribution over existing, legacy, a.c. power supply cabling;
  (ii) LV (380/400 V) d.c. systems under consideration for the powering of equipment in data centres.

(d) the growth in availability of sources that provide d.c. power (for example, wind turbines, solar photo-voltaics, fuel cells and associated hardware).

(e) the growth in availability of low cost and efficient d.c. voltage conversion devices.

3.2 Service integration

Information technology (IT) applications of LV d.c. using premises telecommunications cabling include the powering of devices such as wireless access points and distributed antenna/in-building wireless systems.

The non-IT services supported by LV d.c. using premises telecommunications cabling include primary and emergency lighting, portable device charging, environmental control (heating ventilation and air-conditioning (HVAC)), monitoring and control, alarms, building management/control systems including security systems (surveillance cameras), access control and digital signage.

The integration of these services at the end-device level or via their control systems provides substantial benefits for building and energy management.

3.3 Energy efficiency

Correctly designed centralised distribution of power using LV d.c. increases the overall energy efficiency of both the operation and battery charging of the growing number of d.c.-powered end devices (by removing the many locally installed a.c./d.c. converters and replacing them with centralised, larger and more efficient devices). This has to be balanced to some degree by considerations of the power dissipation in the cabling infrastructure used to supply the LV d.c. solution.

It is also recognised that the increased granularity of control of d.c.-powered end devices, such as lighting systems, will reduce their energy consumption – particularly when more focussed and effective use of those systems is enhanced via service integration.

Recognised standards of d.c. power distribution over telecommunications cabling

4.1 Cabling

The telecommunications cabling addressed by this Section is that of the three recognised premises-specific cabling standards that are:

(a)   the BS EN 50173 series;
(b)   ISO/IEC 11801 and associated premises-specific standards (soon be republished as the ISO/IEC 11801 series);
(c)   the ANSI/TIA-568 series of standards and their associated premises-specific standards.

While the BS EN and ISO/IEC standards contain options for freedom of implementation, they all include reference implementations that use 4-pair (8 conductor) balanced cables and connecting hardware of a given Category:

(d)   BS EN 50173 and ISO/IEC 11801 standards contain requirements for Categories 5, 6, 6_A, 7 and 7_A;
(e)   ANSI/TIA-568 standards contain requirements for Categories 5e, (essentially equivalent to Category 5 above), 6 and 6A (broadly equivalent to Category 6_A above).

4.2 Systems

4.2.1 General

The recognised standard-setting body that produces standards for remote powering via LV d.c. using the cabling discussed in 4.1 is IEEE – specifically, their IEEE 802.3 series of standards:

(a)   IEEE 802.3af: defined a power supply over 2 pairs (4 conductors) with 350 mA per pair (i_c = 175 mA);
(b)   IEEE 802.3at: defined two solutions in which the IEEE 802.3af solution was re-defined as IEEE 802.3at Type 1 (known as PoE);
(c)   IEEE 802.3at Type 2 (known as PoEP): defined a power supply over 2 pairs (4 conductors) with 600 mA per pair (i_c = 300 mA).

A new IEEE specification (expected to be defined as IEEE 802.3bt) is being proposed to use all four pairs with currents per pair of up to 1,000 mA (i_c = 500 mA). Although final decisions on this have not been taken it is clear that the cabling of 4.1 will be subject to increased power loads. No proposals are being made to the cabling component specifications (to support backwards compatibility) and careful consideration is therefore being given to planning and operational requirements and recommendations.

In all these standards, a four-pair cable will supply a load using either 2 or 4 pairs as shown in Figure 2.

A 2-pair powering solution supplies a current to the load using the two conductors of a balanced pair cable element and returns that current from the load using the two conductors of another pair.

A 4-pair powering solution supplies a current to the load using the four conductors of two balanced pair cable elements and returns that current from the load using the four conductors of the two other pairs.

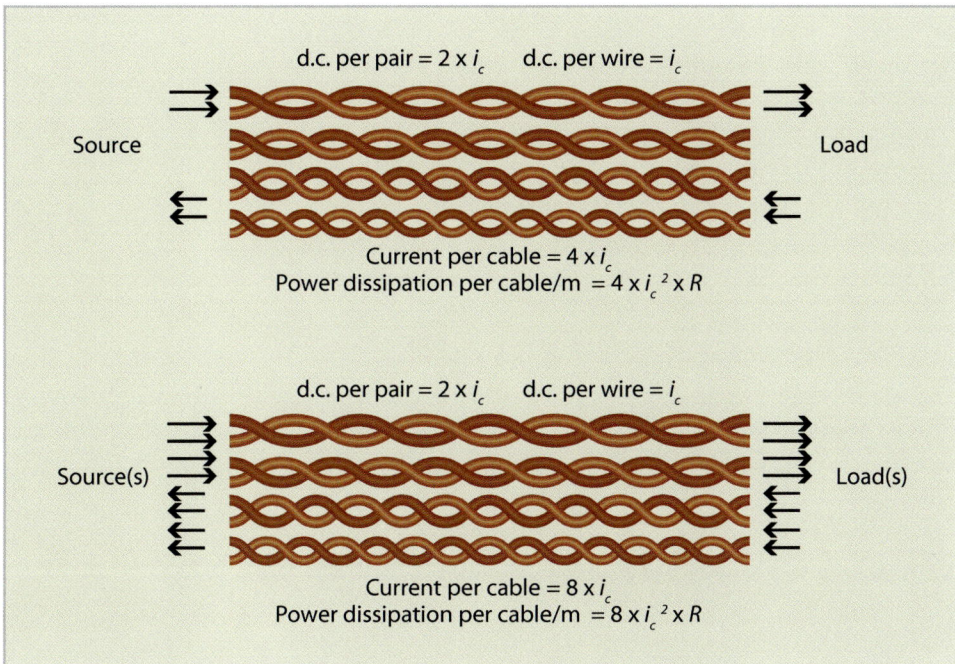

Figure 2: Basic current feeding schematic

d.c. per pair = $2 \times i_c$ d.c. per wire = i_c

Source   Load

Current per cable = $4 \times i_c$
Power dissipation per cable/m = $4 \times i_c^2 \times R$

d.c. per pair = $2 \times i_c$ d.c. per wire = i_c

Source(s)   Load(s)

Current per cable = $8 \times i_c$
Power dissipation per cable/m = $8 \times i_c^2 \times R$

4.2.2 Power Sourcing Equipment (PSE) and Powered Device (PD) detection and classification protocols

4.2.2.1 IEEE 802.3af/802.3at Type 1 and IEEE 802.3at Type 2

To avoid damaging legacy data equipment that does not expect to see d.c. voltage, the IEEE specifications define a mechanism that determines when the PSE may apply and remove power, which is summarised as follows:

(a)   in any operational state, the PSE shall not supply operating power to the PD until the PSE has successfully detected a PD requesting power;

(b)   the PSE probes the link segment in order to detect a valid PD detection signature;

(c)   a PSE shall accept as a valid signature a link segment with the following PD detection signature electrical characteristics:

     (i)   resistance in the range 19.0 kΩ – 26.5 kΩ;

     (ii)   parallel capacitance ≤ 0.15 µF;

(d)   a PSE shall reject resistances above 33 kΩ and below 15 kΩ.

The ability of the PSE to query the PD in order to determine the power requirement of that PD is called 'classification'. The interrogation and power classification function is intended to establish mutual identification and is intended for use with advanced features such as power management.

Mutual identification is the mechanism that allows a Type 2 PD to differentiate Type 1 PSEs from Type 2 PSEs. Mutual identification will also allow Type 2 PSEs to differentiate between Type 1 and Type 2 PDs. PSEs and PDs that do not implement classification can only perform as Type 1 devices.

There are two forms of classifications: Physical Layer classification and Data Link Layer classification.

Physical Layer classification occurs before a PSE supplies power to a PD when the PSE asserts a voltage and the PD responds with a current representing a limited number of power classifications. Based on the response, the minimum power level is set at the output of the PSE. Physical Layer classification encompasses two methods known as 1-Event Physical Layer classification and 2-Event Physical Layer classification.

With the Data Link Layer classification, the PSE and PD communicate using the Data Link Layer protocol integrated with the Link Layer Discovery Protocol (LLDP) after the data link is established. The Data Link Layer classification has finer power resolution and the ability for the PSE and PD to participate in dynamic power allocation wherein allocated power to the PD may change one or more times during PD operation.

Data Link Layer classification takes precedence over Physical Layer classification.

In IEEE 802.3af, the classification step is optional: if a PSE chooses not to classify a PD, it must assume that the PD is a 13 W (IEEE 802.3af) Type 1 device.

4.2.2.2 IEEE 802.3bt

At the time of publication of this Code of Practice the final specification of IEEE 802.3bt has not been published.

4.3 Thermal considerations

4.3.1 General

The cables and connecting hardware of 4.1 were not originally intended to carry current other than signal current. As a result, the delivery of LV d.c. over those components raises a number of risks:

(a)   increases in the operating temperature of the cables (exceeding their specified operating temperature);
(b)   damage to connecting hardware contacts where mating and de-mating occurs while the power supply current is flowing;
(c)   the associated increase of channel attenuation/insertion loss, due to the increased temperature of installed cables, which, unless balanced by reduced installed lengths, will have a negative effect on channel ACR (and may be associated with increased system bit error rates).

In order to address these risks, a number of documents have been, or are being, prepared by the responsible standard-setting bodies.

4.3.2 Published standards

4.3.2.1 ISO/IEC TR 29125:2010

ISO/IEC TR 29125:2010 contains information on temperature rises found at the centre of groups (known as 'bundles') of cables of Category 5 and above and for $i_c \leq 360$ mA.

This information was used during the development of IEEE 802.3at to define the limits of power delivered.

ISO/IEC TR 29125:2010 suggests that cables of a higher Category produce lower temperature increases for a given current and bundle size (N). However, all Categories of cable have the same maximum d.c. resistance specification based on a minimum conductor diameter and so, in practice, the actual conductor diameter and the d.c. resistance per unit length of the conductor may have more influence than the cable Category.

ISO/IEC TR 29125:2010 suggests that the temperature rise increases as both the current per pair increases and as the bundle size increases. However, external surfaces of the bundles were in open air, allowing cooling by radiation and convection. This is not representative of the full range of pathway systems and when bundles are installed in unventilated cable management systems or in insulating materials the temperature rise will be higher.

4.3.2.2 BS EN 50173-1:2011 and ISO/IEC 11801 Ed.2.2:2011

BS EN 50173-1:2011 specifies the d.c. loop resistance requirements of generic cabling channel and links.

BS EN 50173-1:2011 specifies the current carrying capacity requirements of generic cabling channels that are linked to the 10 °C temperature rise suggested in ISO/IEC 29125: 2010 at the centre of a bundle of 100 cables (N=100) for i_c = 300 mA (all conductors powered). Additionally it is stated that:

(a)   relevant application standards and manufacturers' instructions shall be consulted with reference to safety aspects of power feeding;

(b)   care shall be taken when using multi-unit or bundled cables due to the possible rise of temperature within the cabling components that may degrade channel performance.

See 4.3.4 for further detail about damage to components.

4.3.2.3 BS EN 50173-6:2014

BS EN 50173-6 specifies generic cabling for distributed building services that is expected to be exploited by a wide range of remote powering solutions including those of IEEE 802.3 and other proprietary products.

4.3.2.4 BS PD CLC TR 50174-99-1

CLC TR 50174-99-1 has been developed to address the planning and installation issues associated with the increased use of cables and connecting hardware of the EN 50173 series standards for the distribution of both standardised and proprietary implementation of LV d.c. It also contains testing protocols and mathematical modelling tools to support its contents.

4.3.3 Future developments

4.3.3.1 EN 50173-1

A revision of EN 50173-1:2011 is expected to modify the requirements for both d.c. loop resistance and current carrying capacity in order to better control the application and outcomes of remote powering over generic cabling.

The modifications to the d.c. loop resistance requirements are intended to prevent the installation of fixed cables that do not conform to the referenced cable specifications. The modified requirements are expected to include that:

(a)   the channel requirements will be augmented by a 'met by design' requirement for the d.c. loop resistance per unit length;

(b)   the higher performance channels may feature more stringent requirements;

(c)   the link requirements, which are already length dependent, will be modified as per any changes to the channel limits mentioned above to define requirements at 20 °C (to which measured values are required to be corrected).

The existing current carrying capacity requirements are expected to be deleted and to be replaced with:

(d)   a restriction of current carrying capacity to 0.75 A per conductor under continuous operation (as specified in Annex D of EN 50173-1:2011);

(e)   a warning that this is not the current carrying capacity if mating or de-mating under load and not a guide for application support;

(f)   a requirement that any equipment connected to the channels or applications operating over them shall be fitted with over-current protection not exceeding 0.75 A per conductor.

4.3.3.2 ISO/IEC 11801-1
The structure of ISO/IEC 11801 and its associated premises-specific cabling standards is being reorganised and its basic transmission requirements will be republished as ISO/IEC 11801-1. It is expected that most, if not all, of the proposed changes to EN 50173-1 will be incorporated during that revision phase.

4.3.3.3 ANSI/TIA-568 series
The standards in the ANSI/TIA-568-C series are being revised at this time and it is expected that the issues raised during consultation of the proposed changes to EN 50173-1 will be addressed during that revision phase.

4.3.4 Risk of damage to components

4.3.4.1 Fault conditions
The remote powering solutions of IEEE 802.3 at prevent damage to the cabling components under fault conditions (short- or open-circuits) by automatically removing the power supply if the resistance between the conductors that provide the power moves outside a specified range. This is described further in 4.2.2.

4.3.4.2 Cables
Cables specified in the reference implementations of the standards detailed in 4.1 are required to operate at temperatures of ≤ 60 °C.

Consequently, the selection of cables is critical if the cable design- and installation-related factors indicate that cable temperature will exceed this value due to the current levels specified.

Cable suppliers may provide products that have specified maximum operating temperatures in excess of this requirement. The current applied to cables, in conjunction with the intended ambient temperature, should not cause their operating temperature to rise above that specified by the cable supplier.

4.3.4.3 Connecting hardware

The connecting hardware specified in the standards detailed in 4.1 is required to support a continuous operating current of ≥ 0.75 A per conductor at 60 °C. At this time, it is not considered likely that standards-based LV d.c. solutions will exceed this value. Connecting hardware suppliers may provide products that have specified maximum operating currents in excess of this requirement.

A more relevant concern is the risk of damage to the contacts of, or the circuitry embedded within, the connecting hardware when those contacts de-mated under load (i.e. when carrying current). BS EN 60512-9-3 specifies an endurance test for connecting hardware under conditions of mating or de-mating under a disconnection load of 600 mA per conductor.

NOTE: 300 mA is the operational current that under conditions of de-mating can cause 600 mA to flow through the last connector contact to de-mate.

BS EN 60512-9-3 refers to BS EN 60512-99-001, which defines an associated test schedule. This schedule requires connecting hardware to meet the required performance after 100 cycles (as compared to 750 cycles under conditions of mating or de-mating without load).

NOTE: Conformance to BS EN 60512-9-3 and BS EN 60512-99-001 is not an automatic requirement of any connecting hardware interface specification.

The selection of connecting hardware is therefore critical if i_c is greater than 300 mA and/or operating practices are not able to minimise de-mating under load.

4.3.4.4 Design options

Connecting hardware with stated compliance with the standards addressing performance during de-mating under load should be specified.

Further measures should be considered to avoid damage to either third party equipment or to installed connections that may be subject to repeated disconnection under load including:

(a)   when designing all cabling in a new premises the implementation of Consolidation Points (CP) of BS EN 50173-2 or Service Concentration Points (SCP) of BS EN 50173-6 should be considered. These allow the repair of the Telecommunications Outlet (TO) or Service Outlet (SO) respectively by the replacement of CP or SCP cord (as opposed to the re-termination or re-installation of cable at the TO or SO).

(b)   'Bring Your Own Device' or equivalent concepts need to be implemented with clear instructions that all disconnections should be made at the TO/SO (for example, at the wall socket) and not at the equipment (i.e. no liability for damage to the RJ45 port of any third party equipment will be accepted).

4.4   Practical impact of thermal effects

4.4.1   Transmission performance

The channel requirements of all the cabling standards referenced in 4.1 are temperature independent, i.e. they are required to be met at the operating temperature of the channel.

However, any temperature rise due to power feeding in combination with the ambient temperature produces an increase in attenuation/insertion loss of the installed cabling.

The attenuation increase is assumed be, for operating temperatures above 20 °C, 0.2 % per °C for screened cables, 0.4 % per °C (20 °C to 40 °C) and 0.6 % per °C (> 40 °C to 60 °C) for unscreened cables. In order for the cabling performance to meet a given Class, the length of the channel has to be reduced accordingly.

4.4.2 Reduction in transmission distance

Table 1 (taken from CLC TR 50174-99-1) provides general guidance on the reduction of channel and link lengths, which can be applied to all cabling using cables specified in the reference implementations of the standards (i.e. independent of cable construction).

It should be noted that testing the transmission parameters of the installed infrastructure immediately following its installation will not show any impact of increased ambient temperature as the cables will not be 'under load', so planning of cable lengths has to take into account the predicted temperature increases.

Increased operating temperatures may reduce the length over which an application can be supported.

Groups or bundles of cables may contain a mix of applications – some of which have no remote powering content and others within which the levels of remote powering may be continuous or variable. As a result, the combined thermal impact should be considered for all the applications to be supported. To maximise the continuity of application support, the predicted thermal impact should be documented and appropriate planning should be employed when selecting components, installation environment and installation techniques.

	Total length of cords (m)		
	10	15	20
Temperature (°C)	Channel length (m)		
20	100	98	95
25	98	96	93
30	97	94	91
35	95	92	89
40	93	90	87
45	90	87	85
50	86	84	82
55	83	81	79
60	80	78	76

Table 1 – The impact of temperature on channel length

4.4.3 Power dissipation

As a direct result of d.c. transmission with constant current, power is dissipated along the length of the cables. Table 2 provides details of the power dissipation of various powering solutions. The relatively small levels of dissipation associated with IEEE 802.3at solutions (≤ 34 mW/m) rise rapidly with the forecast implementations now being considered. It should also be realised that the values shown are for a metre length

© The Institution of Engineering and Technology

of one Category 5 cable in accordance with the standards of 4.1 and that the total dissipation for long lengths of groups or bundles of cables may represent a potential design and/or operational challenge to HVAC systems in the areas of distribution.

	Power dissipation along cable length (mW/m)	
i_c (mA)	2-pair powering	4-pair powering
175	12	–
300	34	68
500	–	190
750	–	428

4.4.4 Temperature rises

4.4.4.1 Fundamental assumptions

Experiments reviewed during the development of CLC TR 50174-99-1 have shown that there is minimal difference in the temperature rise between groups or bundles of cables that are randomly assembled and those that are very carefully organised.

The mathematical model of CLC TR 50174-99-1 indicates that for a given cable type and installation environment:

(a)   the temperature rise in a group or bundle of cables where $N = 37$ is 20 % higher than that of $N = 24$ – a 20 % design margin is considered appropriate and therefore predictions for $N = 37$ are applied to groups or bundles of cables where $N = 24$;

(b)   for $N \leq 24$ the temperature rise at the outer surface of the group or bundle of cables is generally $\geq$ 80 % of the total temperature rise of the centre of the group or bundle – for this reason it is assumed that all cables experience the same temperature rise (i.e. that of the centre cable).

NOTE: For $N > 37$, the external bundle temperature falls below the $\geq$ 80 % figure but for the purposes of data handling the same assumption is applied.

4.4.4.2 Predictions

Figure 3 shows the predictions of the mathematical model of CLC TR 50174-99-1 for the temperature increase profiles for a variety of cables and installation environments. i_c is restricted to 0.75 A in accordance with 4.3.2.2 but the trends for higher currents are obvious.

NOTE: the mathematical model used includes the 'iterated' increases in conductor resistance as temperatures rise.

The predictions are in close agreement with data presented in ISO/IEC TR 29125:2010 for the range of cables considered for $N = 37$ and for $i_c = 300$ mA (all conductors powered).

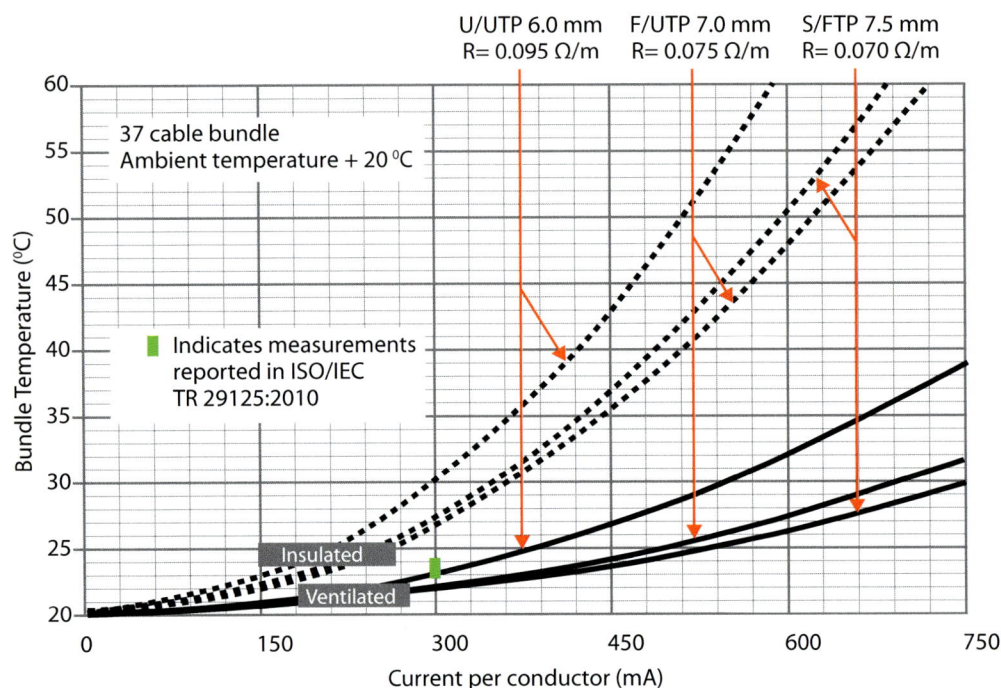

Figure 3: Range of temperature rises

4.4.4.3 Predictions

Typical installations would consist of groups or bundles of cables routed in a variety of pathways and pathway systems within which the temperature rises would vary from place to place along the length of the route. As a result, detailed operational guidance is not possible and any recommendations have to be based on generalised assumptions.

However, the rapid increase of temperature as currents are increased, and the even more dramatic increase as the ventilation around the bundles is reduced, justifies the need to design for and/or control the current injected into groups or bundles of cables.

There are two separate aspects to be considered:

(a) unventilated or insulated cables of extended length when subjected to relatively low values of i_c, even for low values of N, generate temperature rises that exceed the generic maximum operating temperatures of the cables – potentially resulting in damage to the cable construction and rendering it unsuitable for transmission of at least some of the applications listed in Annex F of EN 50173-1:2011; and

   NOTE: The effect on short lengths of groups or bundles of cable passing through fire barriers is for further study.

(b) ventilated cables of extended length when subjected to higher values of i_c than those used by IEEE 802.3at can, even for low values of N, produce temperature increases that would result in reductions of viable channel length as shown in Table 1.

Groups or bundles of cables that are installed in pathways or pathway systems that are subject to natural or forced ventilation produce the lowest temperature rises.

Groups or bundles of cables that are installed in pathways or pathway systems that are not subject to natural or forced ventilation (such as conduit or trunking) produce greater temperature rises by a multiplier of ~2.

Groups or bundles of cables that are installed in insulated pathways or pathway systems produce significantly higher temperature rises than those that are subjected to natural or forced ventilation. The multiplier depends upon the degree of installation but values of 6 have been observed.

4.5 Controlling injected power

The applications of IEEE and many other proprietary remote powering systems typically operate at voltages not exceeding 60 V d.c. delivering power using some or all of the pairs within a cable. Table 3 indicates the total current and total power injected into a group or bundle of cables where N = 24, which would result in specified temperature rises based on the mathematical model of CLC TR 50174-99-1.

NOTE: the mathematical model used includes the 'iterated' increases in conductor resistance as temperatures rise.

ΔT(°C)	Ventilated conditions			Insulated conditions		
	i_c(A)	Total group/ bundle current (A)[1]	Injected group/bundle power (kW)[2]	i_c(A)	Total group/ bundle current (A)[1]	Injected group/bundle power (kW)[2]
5	0.39	74.9	2.1	0.20	38.4	1.1
10	0.55	105.6	3.0	0.29	55.7	1.6
15				0.35	69.1	2.0
20				0.40	78.7	2.2
25	> 0.75[3]	> 144.0[3]	> 4.3[3]	0.44	88.3	2.5
30				0.48	97.9	2.8
35				0.51	105.6	3.0
40				0.54	113.3	3.2

[1] $8 \times N \times i_c$
[2] $60 \times 4 \times i_c \times N$
[3] i_c would exceed 0.75A and would be outside the limit specified by standards

NOTE: the above values assume the following modelling conditions:
(a) U/UTP cables
(b) R = 0.095 Ω/m
(c) d = 0.005 m

Table 3 – Temperature rise and injected power (U/UTP cables)

4.6 Recommendations

4.6.1 Selection and erection of equipment (installation)

4.6.1.1 Components
Cables having conductors of the lowest practicable d.c. resistance should be selected as they will dissipate lower levels of power for a given value of i_c.

Cables of the highest practicable Category should be selected as they will exhibit better insertion loss/attenuation performance and will thereby maintain the destined performance to higher temperatures.

Cables with the highest practicable operating temperature should be selected.

© The Institution of Engineering and Technology

Connecting hardware with stated compliance with the standards addressing performance during de-mating under load (see 4.3.4.3) should be specified.

4.6.1.2 Link lengths

The maximum lengths of installed cable may have to be reduced to take into account the increases in insertion loss/attenuation produced by the increased temperature association with LV d.c. distribution within the group or bundles of cables.

This should be taken into account during any cabling acceptance tests.

4.6.1.3 Installation

Cabling in accordance with the BS EN 50173 series of standards requires the planning and installation of cabling to be in accordance with the BS EN 50174 series of standards, which requires the application of the manufacturer's/supplier's instructions.

In addition to those requirements the following should be applied:

(a)   bundles of $N \leq 24$ should be installed;
(b)   air gaps should be provided between groups or bundles of cables to allow for ventilation.

Pathway systems and installation methods should be selected to maximise the opportunity for the outer surfaces of cables to be cooled by surrounding airflow whilst maintaining the requirements for segregation of Section 6 of EN 50174-2 in relation to electromagnetic interference. An example is shown in Figure 4.

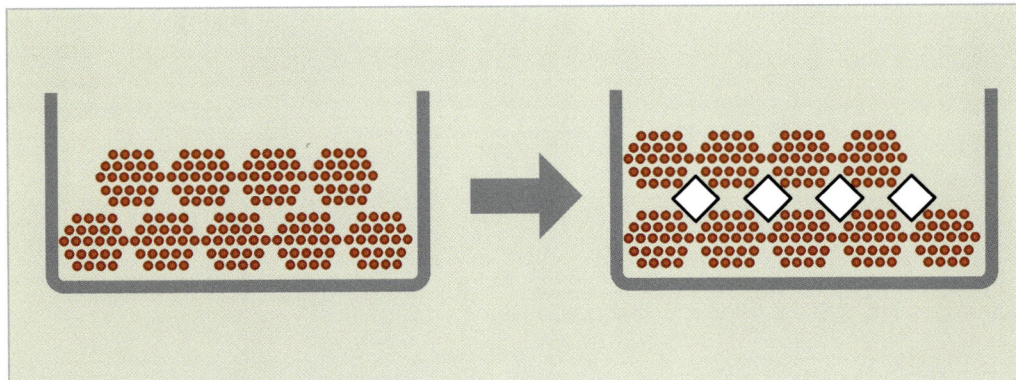

Figure 4: Example of airflow improvement in pathway system installation

4.6.2   Legacy cabling installations

In legacy installations it is important to avoid both the increases in attenuation associated with increases in temperature and the risk of damage to installed cabling.

Where cables are run in non-ventilated spaces or in containment that provides inherent thermal insulation properties, consideration should be given to installing mechanical ventilation.

4.6.3   Documentation

The lengths and types of pathway and pathway systems (for example, lengths of insulated pathway) and the installation configuration (for example, presence of air gaps) of cable bundles within them should be documented for future reference.

Accurate records should be maintained. In high density cabling the installation and operational complexity for the administration of the cabling infrastructure, in accordance with EN 50174, should be increased one level.

4.6.4 Application of power

All power supply equipment and end-devices connected to the cabling should conform to the relevant IEEE specification.

Power may be injected at the data source equipment (for example, at the switch) or via separate mid-span power insertion equipment.

The maximum specified power that may be injected into each cable bundle should be recorded for mapping against Table 3.

Where deemed necessary, the power injected should be amortised across cable bundles to minimise effective loading.

Accurate records should be maintained.

4.7 Cabling acceptance and trouble-shooting testing

4.7.1 General

The majority of acceptance testing of installed cabling against the requirements of a given Class (according to BS EN 50173-1 or ISO/IEC 11801) or a given Category (in accordance with ANSI/TIA-568 standards) is undertaken when the ambient temperature is at or below normal human occupancy conditions (20 °C or below).

4.7.2 Insertion loss/attenuation

The test equipment assesses the insertion loss/attenuation of each pair against a specified value (independent of length) despite a length dependency being included in the formulae that produces that limit. Consideration should be given to the validity of the measurement if it is made under unpowered conditions and may perhaps need to be reassessed against a lower limit that is consistent with the maximum length and the applicable temperature.

4.7.3 d.c. loop resistance

Developments in remote powering have moved faster than the installed cabling requirements of the main structured cabling standards. As a result, the requirements for d.c. loop resistance and its measurement are subject to considerable shortfalls. These shortfalls will be addressed in future revisions of the cabling standards as described in 4.3.3. However, in the interim they are detailed below.

The requirements for the d.c. loop resistance in all the cabling standards referenced in 4.1 are based on cables of a d.c. resistance per conductor of 0.095 Ω/m operating at 60 °C. This results in virtually any cable (even cables with much higher values of d.c. resistance per conductor) when measured at a lower ambient temperature being able to pass the requirements.

In addition, the ANSI/TIA-568-C standards do not require high values of d.c. loop resistance to be reported during standard-based test regimes.

The d.c. loop resistance requirements for installed cabling links (for example, outlet to outlet) designed in accordance with the EN 50173 series, and their ISO/IEC equivalents, are length dependent. Unfortunately, most test equipment only applies a specified value (based upon a 90 m length). It is therefore possible for short lengths to comprise cables of high resistance for which no indication is made within the test result.

Consideration should be given to the validity of the measurement if it is made under unpowered conditions and may perhaps need to be reassessed against a lower limit that is consistent with the maximum length, the applicable temperature and the specified (rather than standardised) d.c. resistance of the installed cable.

Proprietary d.c. power distribution over telecommunications cabling

5.1 Cabling

The cabling addressed by this Section is that described in 4.1. However, the remote powering applications of this Section include solutions that deliver values of conductor current (i_c) exceeding those supported by minimally compliant components of the Categories described in the standards of 4.1. For this reason, this Section contains data for implementations in addition to those of Section 4.

5.2 Systems

The proprietary d.c. power distribution systems of this Section are assumed to use maximum voltage levels that comply with the SELV requirements of BS EN 60950-1.

NOTE: Telecommunications providers' applications, such as ISDN and ADSL, that feature voltage levels in excess of the SELV requirements of BS EN 60950-1 are addressed in 6.3.

A four-pair cable will power a load using either 2 or 4 pairs as shown in Figure 2.

A 2-pair powering solution supplies a current to the load using the two conductors of a balanced pair cable element and returns that current from the load using the two conductors of another pair.

A 4-pair powering solution supplies a current to the load using the four conductors of two balanced pair cable elements and returns that current from the load using the four conductors of two other pairs.

However, the power delivery systems of this Section are not assumed to feature any of the power delivery controls (methodologies or current controls) of the standards-compliant implementations of Section 4.

5.3 Thermal considerations

5.3.1 General

A number of proprietary remote powering solutions are available that deliver power levels significantly in excess of IEEE 802.3at Type 2 – and, in some cases, in excess of those being considered by IEEE within the 802.3bt project. Caution should be exercised when selecting any solution that claims IEEE compliance whilst offering enhanced levels of power delivery, as the two claims are contradictory.

At the time of writing, the highest power delivery solution that references the telecommunications cabling described in Section 4 uses $i_c = 1$ A and uses all pairs to provide up to 232 W injected power per cable.

While such solutions operate outside the boundaries defined by the minimally-compliant cabling components specified in the reference implementations of the standards detailed in 4.1, it is possible for carefully selected components to support those solutions provided that the appropriate safeguards of this Section are employed.

However, even if the proprietary solution provides power at levels consistent with the published IEEE standards, it cannot be assumed that they meet the other requirements of the IEEE specifications summarised in 4.2.2.

5.3.2 Risk of damage to components

5.3.2.1 System specification

Proprietary solutions should only be considered if the following information is clearly and unambiguously specified:

(a)   the voltage supplied across the pairs providing the power supply;

(b)   the maximum current supplied per conductor (i_c);

(c)   the number of pairs used (2 or 4);

(d)   the process adopted by the PSE following the connection of a power-capable circuit (i.e. a cabling channel without a PD attached) or with a PD attached that is not designed for use with the solution;

(e)   the process adopted by the PSE following the connection of a PD designed for use with the system;

(f)   the process adopted by the PSE following the breaking of a circuit or the disconnection a PD.

5.3.2.2 Fault conditions

Any connected equipment should be fitted with over-current protection such that the cabling and connecting hardware are not subject to conditions exceeding their operating specification (in accordance with BS 7671).

Based upon the information provided by the solution supplier, it should be determined if any significant risk for the intended application exists by the attachment of PDs, which are not designed for use with the specified PSE.

5.3.2.3 Cables

For cables specified in the reference implementations of the standards detailed in 4.1, the maximum operating temperature is specified as $\geq 60\,°C$. Consequently, the selection of cables is critical if the cable design – and installation-related factors – indicate that cable temperature will exceed this value due to the current levels to be employed.

Cable suppliers may provide products that have specified maximum operating temperatures in excess of this requirement. The current applied to cables, in conjunction with the intended ambient temperature, should not cause their operating temperature to rise above that specified by the cable supplier.

5.3.2.4 Connecting hardware

The connecting hardware specified in the standards detailed in 4.1 is required to support a continuous operating current of ≥ 0.75 A per conductor at $60\,°C$. Connecting hardware suppliers may provide products that have specified maximum operating currents in excess of this requirement. The current applied to each conductor of the connecting hardware should not exceed the value specified by the connecting hardware supplier.

A more relevant concern is the risk of damage to the contacts of, or the circuitry embedded within, the connecting hardware when those contacts de-mated under load (i.e. when carrying current). BS EN 60512-9-3 specifies an endurance test for connecting hardware under conditions of mating or de-mating under a disconnection load of 600 mA per conductor.

NOTE: 300 mA is the operational current that under conditions of de-mating can cause 600 mA to flow through the last connector contact to de-mate.

BS EN 60512-9-3 refers to BS EN 60512-99-001 which defines an associated test schedule. This schedule requires connecting hardware to meet the required performance after 100 cycles (as compared to 750 cycles under conditions of mating or de-mating without load).

NOTE: BS EN 60512-9-3 and BS EN 60512-99-001 is not an automatic inclusion in any connecting hardware interface specification.

The selection of connecting hardware is therefore critical if i_c is greater than 300 mA and/or operating practices are not able to minimise de-mating under load.

5.4   Practical impact of thermal effects

5.4.1   Transmission performance

See 4.4.1.

5.4.2   Reduction in transmission distance

See 4.4.2.

5.4.3   Power dissipation

As a direct result of d.c. transmission with constant current, power is dissipated along the length of the cables. Table 2 provides details of the power dissipation of various powering solutions up to the maximum current levels specified for connecting hardware in the cabling standards of 4.1. Table 4 extends the contents of that table to higher currents seen with proprietary systems of LV d.c. power distribution.

It should be realised that the values shown are for a metre length of one Category 5 cable in accordance with the standards of 4.1 and that the total dissipation for long lengths of groups or bundles of cables may represent a potential design and/or operational challenge to HVAC systems in the areas of distribution.

i_c (mA)	Power dissipation along cable length (mW/m)
	4-pair powering
1 000	760
1 500	1 710

Table 4 – Power dissipation for $i_c \leq 1\ 500$ mA, $R = 0.095\ \Omega/m$

5.4.4 Temperature rises

Levels of i_c in excess of those supported by minimally-compliant cabling components specified in the reference implementations of the standards detailed in 4.1 will raise the temperature increases seen within a cable bundle of a given value of N.

The consideration given to bundle sizes and the provision of ventilation becomes increasingly important as the mathematical model of CLC TR 50174-99-1 predicts that, for a given value of N, and a common installation environment, the temperature rise is proportional to i_c^2.

5.5 Controlling injected power

Proprietary remote powering solutions complying with 5.2 to be operated over telecommunications cabling of 5.1 should be restricted to those that:

(a)   adopt values of $i_c \le$ the current carrying capacity of the installed connecting hardware;

(b)   have considered and addressed the impact of de-mating of connections under load;

(c)   are fitted with over-current protection at a level $\le$ current carrying capacity of the installed connecting hardware.

Table 5 indicates the total current and total power injected into a cable bundle of $N = 24$, which would result in specified temperature rises based on the mathematical model of CLC TR 50174-99-1.

NOTE: the mathematical model used includes the 'iterated' increases in conductor resistance as temperatures rise.

Table 5 – Temperature rise and injected power (U/UTP cables)

DT(°C)	Ventilated conditions			Insulated conditions		
	i_c(A)	Total group/ bundle current (A)[1]	Injected group/bundle power (kW)[2]	i_c(A)	Total group/ bundle current (A)[1]	Injected group/bundle power (kW)[2]
5	See Table 3			See Table 3		
10						
15	0.67	128.6	3.8			
20	0.77	147.8	4.4			
25	0.86	165.1	4.9			
30	0.94	180.5	5.4			
35	1.02	195.8	5.8			
40	1.09	209.3	6.2			

[1] $8 \times N \times i_c$

[2] $60 \times 4 \times i_c \times N$

[3] i_c where supported by the installed connecting hardware

NOTE: the above values assume the following modelling conditions
(a)   U/UTP cables.
(b) $R = 0.095\ \Omega/m$.
(c) $d = 0.005\ m$.

5.6   Recommendations

See 4.6.

5.7   Cabling acceptance and trouble-shooting testing

See 4.7.

Proprietary d.c. power distribution over proprietary cabling

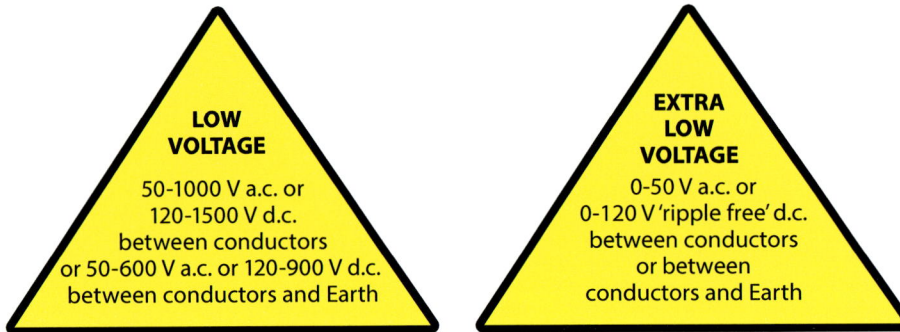

LOW VOLTAGE
50-1000 V a.c. or
120-1500 V d.c.
between conductors
or 50-600 V a.c. or 120-900 V d.c.
between conductors and Earth

EXTRA LOW VOLTAGE
0-50 V a.c. or
0-120 V 'ripple free' d.c.
between conductors
or between
conductors and Earth

6.1   General

6.1.1   Section structure

This Section describes the issues (additional to those of Section 4 and Section 5) that result from the extension of proprietary d.c. power distribution solutions by the use of proprietary cabling as follows:

(a)   6.2 describes the implementation of SELV systems;
(b)   6.3 describes the general implementation of systems operating in the ELV band;
(c)   6.4 describes the implementation of systems operating in the LV band at approximately 400 V d.c. (being proposed for power distribution in areas such as data centres);
(d)   6.5 describes the implementation of systems operating in the LV band between 400 and 1,500 V d.c. (such as those of the subsystems within photovoltaic generation).

Safe working practices are described in HSG85.

6.1.2   Segregation

Segregation in accordance with BS 7671 should be applied between a.c and d.c. cabling for the purposes of safety and protection. Segregation for the purposes of reducing interference of the d.c. power supply is for further study.

Connecting hardware of one power supply solution should not be interchangeable with that used for other power supply solutions. The design of the connecting hardware should maintain the polarity of the circuit.

Where no other form of identification exists, d.c. circuit conductors should be labelled at points of connection to indicate supply polarity.

Tools-only access should be employed to any fitting where screw terminals are used to prevent easy cross-connection to other power supply solutions and labelling of connection points should be employed to distinguish the purpose of individual terminal and the polarity of the circuit.

NOTE: circuits that reverse polarity are not within the scope of this Section.

6.2 Proprietary d.c. powering solutions complying with ELV requirements

6.2.1 Cabling and systems

In this Section, the cabling is assumed to be designed to support the d.c. powering solution and for this reason thermal impact is not addressed.

For the purposes of this Code of Practice it is assumed that these d.c powering solutions employ voltages not exceeding 120 V d.c. and that measures for protection in relation to safety are in accordance with BS 7671.

NOTE: BS 7671 may provide a basis for such a design but there are situations lying outside the assumptions of BS 7671. In such conditions an engineering analysis is required to support the following specific assumptions.

To be specific it is assumed that:

(a)   the cable specification, i.e. the conductor resistance R and associated cable construction parameters, have been selected such that the expected thermal impact of the conductor current (i_c) has been taken into account as part of a system design philosophy;

(b)   the maximum operating temperature of the cables is not exceeded when subjected to the relevant planning installation guides that support the solution;

(c)   the impact of increased operating temperature resulting from the supply of power does not result in a reduction of transmission distance of any data required to be transmitted over the cabling;

(d)   the connecting hardware used (including screw terminals) has been selected to support the specified value of i_c.

As a result, the primary areas of concern are the risk of damage to:

(e)   connecting hardware when de-mating under load;

(f)   cables, connecting hardware and attached equipment (both power source and PD) if subjected to voltages and conductor currents outside the specified operational conditions i.e. by cross-connection to other power supply solutions including mains power supplies.

6.2.2 Fault conditions

Any connected equipment should be fitted with over-current protection such that the cabling and connecting hardware are not subjected to conditions exceeding their operating specification (in accordance with BS 7671).

6.2.3 Implementation

The components and materials used should be in accordance with manufacturer's or supplier's instructions.

The installation and acceptance testing processes employed should be in accordance with manufacturer's or supplier's instructions.

The manufacturer or supplier may define procedures for assessment of installation competence.

6.3 External network operator supplies

6.3.1 General

It is assumed that the only remote powering applications operating in the voltage range 60–120 V d.c. are those of access network service providers. The power sources are external to the premises housing those devices.

The cabling entering the premises is proprietary and is designed specifically to support the voltage and current levels of the service delivered. The PDs are associated with telecommunications transmission systems such as ISDN and ADSL. The use of the cabling of Section 4 and Section 5 to extend these applications to end devices distributed within the premises is not considered to be at an enhanced risk of thermal or other form of damage since its d.c. performance exceeds that of the proprietary cabling entering the premises.

6.3.2 Implementation

See 6.2.3.

6.4 Proprietary LV d.c. powering solutions at approximately 400 V d.c.

There are a variety of proposals for this form of powering in commercial premises. The initial proposals are focussed on the computer rooms of data centres where possible benefits have been suggested as:

(a)   a reduction of capital expenditure due to the elimination of inverters in uninterruptible power supplies and power correction factors in power supplies together with a reduction in copper usage in the distribution systems;
(b)   a reduction of operational expenditure due to improvements in conversion efficiency, distribution losses and reliability coupled with reductions of power dissipation (and associated impact on HVAC);
(c)   improved use of space due to the removal of power conversion equipment;
(d)   improved voltage control;
(e)   simplified connection to alternative energy sources.

The actual levels of benefit attained depend on the design of the existing conventional a.c. distribution system.

Work in this area is at an early stage with reference standards being slow to develop. At this stage the only documents identified are ETSI EN 300 132-3-1 and ETSI EN 301 605.

These implementations represent a considerable hazard if improperly designed, installed or operated. Full and detailed implementation instructions are required from the design stage.

6.5 Proprietary LV d.c. powering solutions in excess of 400 V d.c.

Systems operating above 400 V d.c. represent a considerable hazard if improperly designed, installed or operated. Full and detailed implementation instructions are required from the design stage and may be subject to national or local regulations or application-specific Codes of Practice. For example, see the future IET *Code of Practice for Grid Connected Solar Photovoltaic (PV) Systems*.

Proprietary d.c. power distribution over conventional single-phase a.c. power supply cabling

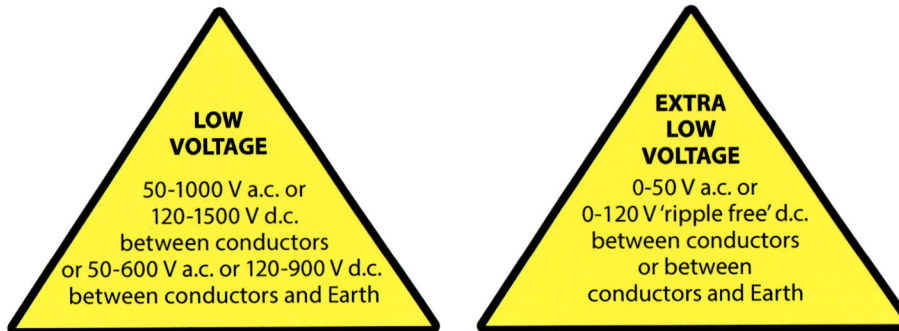

LOW VOLTAGE

50-1000 V a.c. or
120-1500 V d.c.
between conductors
or 50-600 V a.c. or 120-900 V d.c.
between conductors and Earth

EXTRA LOW VOLTAGE

0-50 V a.c. or
0-120 V 'ripple free' d.c.
between conductors
or between
conductors and Earth

7.1   General

7.1.1   Basic principles

For the purposes of this Section, 'conventional single-phase a.c. power supply cabling' is considered to be comprised of LV mains single-phase radial or ring power supply circuits within domestic and commercial premises.

a.c. power distribution circuits containing multiple line and neutral conductors may be converted to distribution of ELV or LV d.c., provided that each d.c. circuit uses two (unipolar d.c.) or three (bipolar d.c.) conductors and does not share the conductors of another d.c. circuit.

Earthing arrangements of circuits should be assessed for:

(a)   suitability to support d.c distribution;
(b)   compliance with current practice as defined in BS 7671.

Existing earthing conductors should be not used as d.c circuit conductors.

7.1.2   Section structure

The structure of this Section is as follows:

(a)   7.2 describes the testing of the a.c. power distribution cabling before initiating any action to convert it to d.c. power distribution;
(b)   7.3 describes the method of determining the capability of the selected a.c. power distribution cabling to support d.c. power distribution;
(c)   7.4 describes the implementation of the d.c. power distribution over the selected a.c. power distribution cabling including migration of consumer units etc;
(d)   7.5 discusses the co-existence of d.c. and a.c. circuits;
(e)   7.6 provides information on labelling;
(f)   7.7 provides additional information on certification of the d.c. power distribution installation.

Safe working practices are outlined in HSE publication HSG85.

7.1.3 Procedures

Conversion of existing single-phase a.c. cabling to support d.c. power distribution is viable and can be conducted in such a way that both safety is maintained and d.c. system performance is warranted.

The overriding assumption within this Code of Practice is that the principal criterion for reusing existing a.c. power supply cabling is to minimise any building work to support the migration to d.c. As such it is assumed that any modification to the existing wiring is limited to, for example:

(a)   disconnection of the a.c. wiring circuit at the a.c. consumer unit and reconnection to a d.c. consumer unit;

(b)   separation and/or re-partitioning of circuits to reduce, for example, circuit resistance;

(c)   replacement of existing a.c. connection points and switchgear with appropriately rated d.c. equivalents to prevent, for example, accidental mixing of a.c. and d.c. loads on the same circuit.
          NOTE: Existing a.c. protection devices may not have appropriate specifications for d.c. operation.

In order to achieve a successful migration it is important that it is conducted using a methodical process. The purpose of this Section is to describe such a process.

The process contains the following steps:

(d)   remove all loads that are unsuitable for connection to the d.c. supply;

(e)   verify the existing wiring in accordance with the current edition of BS 7671;

(f)   assess the performance of the circuit(s) to determine capability limits for d.c. power distribution;

(g)   review and re-design circuit protection to support d.c. power distribution;
          NOTE: Existing a.c. protection devices may not have appropriate specifications for d.c. operation.

(h)   implement the changes necessary to support d.c. power distribution;

(i)   test the converted wiring in accordance with the current edition of BS 7671.

All loads that cannot be connected to the d.c. load shall be removed.

7.1.4 Certification

7.1.4.1 Commercial premises

BS 7671 requires certification by means of an Electrical Installation Certificate for electrical installation work of the type described in this Section, whether it is a new installation or an addition or alteration to an existing installation. This applies to the complete installation i.e. both new and existing wiring.

Issues to be addressed include the following (where the terms used are in the context of BS 7671):

(a)   all installed protection measures should be viable;

(b)   protective bonding and the main earth conductor should be present and correct;

© The Institution of Engineering and Technology

(c)   anything immediately unsafe, noticed during initial survey or installation work, should be rectified or discussed with the customer to agree the required remedial work;

(d)   if the customer does not agree to the remedial work, the contractor should refuse to carry out any new installation work;

(e)   any deviations from BS 7671 that do not pose an immediate safety concern should be noted and the customer informed as appropriate;

(f)   additions to, or alterations of, the installation should be carried out in accordance with BS 7671;

(g)   the extent and limitations of the new work should be clearly defined on the Electrical Installation Certificate;

(h)   the installation should be left in as safe or safer condition as before the new work was performed.

In addition, an Electrical Installation Certificate tailored to the d.c. power distribution cabling should also be completed (see 7.6).

7.1.4.2 Domestic premises

In addition to the certification process described in 7.1.4.1, the re-use of existing a.c. wiring for d.c. power distribution involves significant modifications to the LV a.c. wiring installation and is effectively bonded to the main earthing terminal. Any local, regional or national regulations that apply to such modifications should be applied.

7.1.5   Mixing ELV and LV circuits

With regard to protection:

(a)   Regulation 528.1 of BS 7671 requires that a Band I circuit shall not be contained in the same wiring system as a Band II circuit. Once converted, the ELV d.c. circuits would fall into Band I and other LV d.c. circuits and the existing LV a.c. circuits would fall into Band II. This regulation allows a number of exceptions. The first exception is that it is permitted to mix the circuits if every cable or conductor is insulated for the highest voltage present. Thus Regulation 528.1 permits the mixing of ELV d.c. circuits with LV power circuits such as radial or ring mains power circuits.

(b)   Regulation 515.2 of BS 7671 requires segregation between equipment carrying current of different types or at different voltages to be segregated if they are grouped together in a common assembly. Where this occurs, compliance may require physical replacement of a common assembly with separate assemblies (see 7.4.3).

7.2   Verification of existing a.c. power supply cabling

The purpose of verification is to determine if the existing installation is suitable for conversion to d.c. power distribution, and to ensure that safety is not impaired. For example, the continuity and correct connections of protective earthing and bonding conductors and final circuits (the ring, multiple loop connections, and radial) must be tested before any d.c. installation is added.

The testing criteria (inspection, testing, and certification) are defined in BS 7671.

The verification process should also identify any hidden loads such as in-line fans, the presence of which may not be readily discernible and that could, if left in-circuit, present a safety hazard or could damage either the hidden load or the connected d.c. apparatus.

7.3 Capability assessment

7.3.1 Maximum load and future development

This requires a review of all the demand that needs to be supplied by d.c. power in addition to the consideration of any possible future extensions. Estimation of the maximum demand will depend on the type and diversity of the load and the type of the circuits used within the electrical installation (ELV or LV) in accordance with the following general formula:

$$\text{maximum demand (kW)} = \text{connected load} \times \text{diversity factor}$$

BS 7671 provides suggested diversity factors for domestic and commercial installations. Based on the maximum demand assessment, the current flow in each d.c. circuit of the installation can then be estimated, and the capability of the installation required to support the proposed loads can be verified.

7.3.2 Selection for current carrying capacity (correction factors)

Calculations of current carrying capacity should be undertaken as described in BS 7671.

7.3.3 Voltage drop

The calculation of voltage drop in d.c. circuits is important, particularly where the nominal voltage is low (i.e. ELV applications).

The voltage drop normally depends on the length of circuit (i.e. the maximum length of the radial), the current flowing through the circuit and the cross-sectional area of the conductor.

Calculations of voltage drop should be undertaken as described in BS 7671.

The location of the interruption of ring circuits (see 7.4.1.2) should be selected, where practicable, to minimise the voltage drop in each resulting d.c. circuit.

Information should be available in relation to the voltage and current capability of each circuit or group of circuits.

The equipment to be connected should be selected based upon its voltage and current requirements – which may place restraints on the length of the circuits used.

7.3.4 Quantity of outlets

The number of outlets will be determined by the estimated load and protection constraints.

7.4 Implementation

7.4.1 Simple installations and final circuits

7.4.1.1 d.c.-powered ELV circuit (conductor cross-sectional area ≤ 1,5 mm²)
ELV d.c. circuits using radial circuits with conductors of cross-sectional area of ≤ 1,5mm² can be used for powering low power loads, such as lights, as shown in Figure 5, and other small d.c. loads, such as smoke alarms, burglar alarms etc.

NOTE: The general convention is that the former line conductors provide the positive
d.c. supply and the former neutral conductors provide the negative d.c. supply.

Any earth connections present in switch housings shall be maintained.

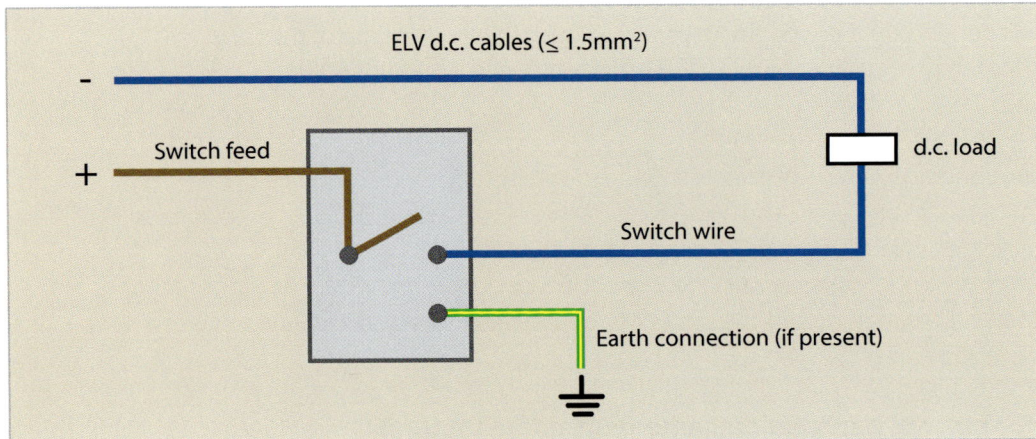

Figure 5: Simple installation of an ELV d.c. powering lighting and controlled by one way switching

7.4.1.2 d.c.-powered LV circuits (conductor cross-sectional area > 1,5mm²)

LV d.c. circuits using radial and ring circuits with conductors of cross-sectional area of > 1,5mm² can be used for powering other loads. Simple radial circuits are as shown in Figure 6.

NOTE: The general convention is that the former line conductors provide the positive
d.c supply and the former neutral conductors provide the negative d.c supply.

Ring circuits should be interrupted to create two radial circuits to simplify load calculations and to improve fault management as shown in Figure 7.

It should be noted that power sockets of the a.c. power supply system should be replaced with sockets that prevent accidental misconnection of d.c. loads to a.c. sockets (and vice versa).

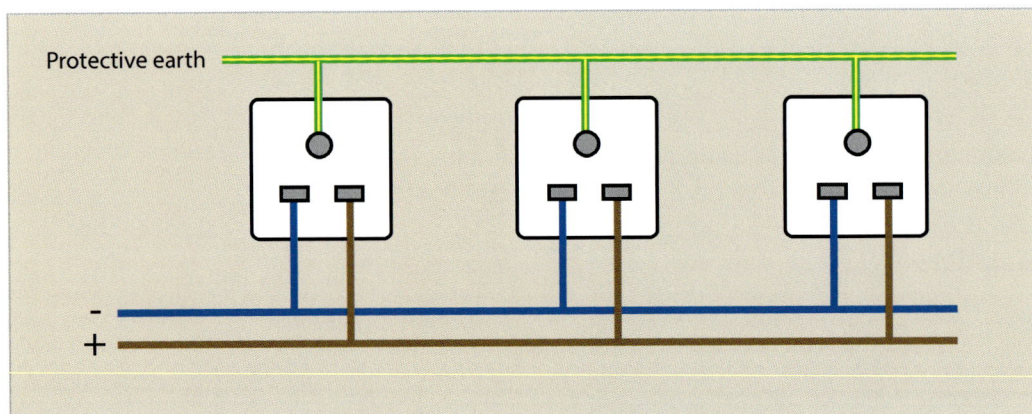

Figure 6: A radial circuit powered by d.c.

NOTE: The provisions of the Plugs and Sockets etc. (Safety) Regulations 1994 apply.

Figure 7:
Modification of an LV a.c. ring circuit to implement d.c. power distribution

Protective earth

Ring circuit interruption

Line
Neutral

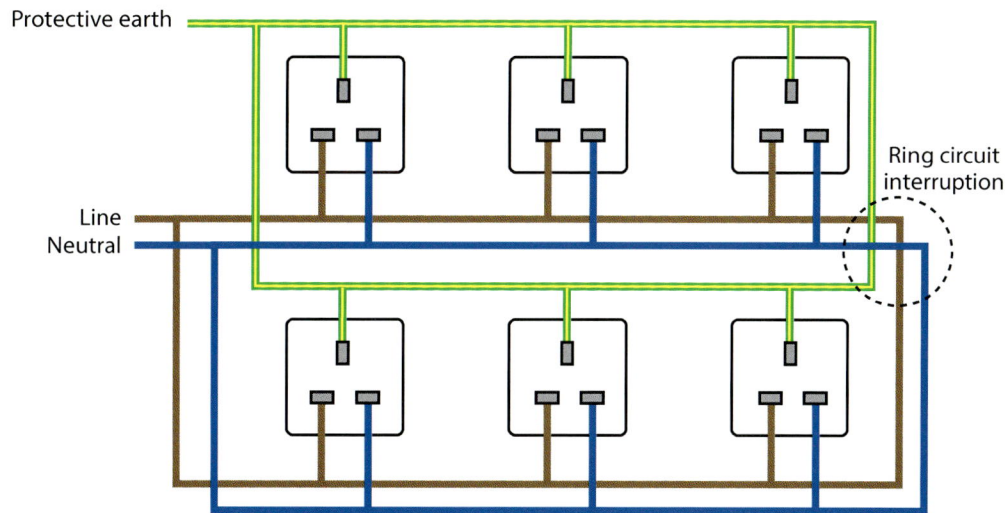

a.c. configuration

↓

d.c. configuration

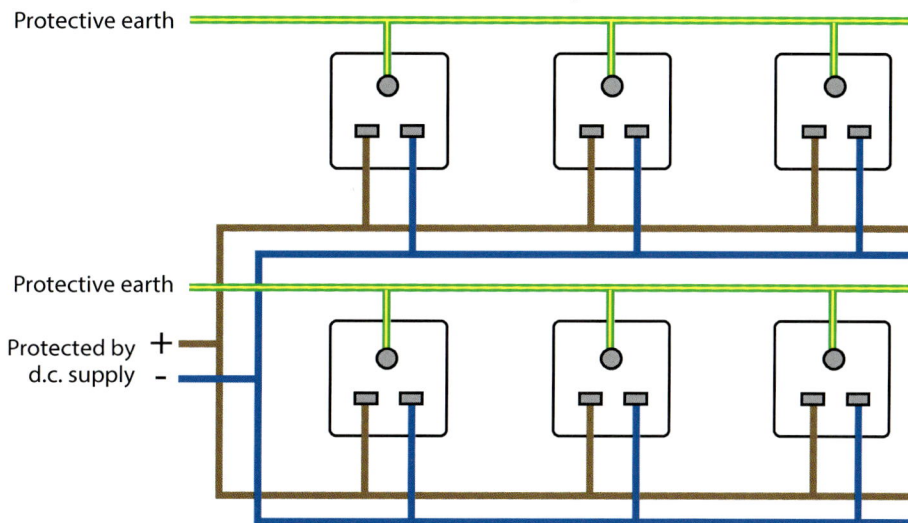

Protective earth

Protective earth

Protected by d.c. supply + −

NOTE: The provisions of the Plugs and sockets etc. (Safety) Regulations 1994 apply.

7.4.2 d.c. power source segregation

Where the d.c. power source is fed from an a.c. source (for example, via an a.c./d.c. converter), suitably rated overcurrent protection devices shall be provided on both the a.c. and the d.c. sides. The segregation of multiple d.c. power sources should be implemented in accordance with BS 7671.

It is recognised that interrupting a d.c. fault current is more difficult than doing so for an a.c. fault current.

NOTE: d.c fault current and voltage waveforms do not have natural zero crossing points, but it is possible to interrupt a d.c. fault if suitably rated fuses or circuit breakers are used, their ratings are adjusted to be applicable to d.c systems, and their operating time complies with the BS 7671 requirements (disconnection times must be sufficiently quick to limit temperature rise in cables).

© The Institution of Engineering and Technology

The d.c. side protection should be designed to be no less safe than the protection that is provided to the remaining a.c. circuits.

7.4.3 Migration issues

Mixing ELV d.c., LV a.c. and LV d.c. circuit junctions in common enclosures, such as consumer units and switches, is considered potentially dangerous and thus necessitates a separate junction box.

As shown in Figure 8, the consumer unit providing d.c. power should be segregated from that which supplies the a.c. power.

Chapter 51 of BS 7671 contains requirements concerning the identification of circuits and use of warning notices. Although there are no specific requirements for notices identifying the presence of LV d.c. circuits, it is suggested that notices are fixed at both the mains supply point and the d.c. control-gear to warn of the presence of mixed a.c. and d.c. circuits. This suggestion is made because the installations will be of an uncommon nature. Further information is provided in 7.4.

NOTE: The general convention is that the former line conductors provide the positive d.c. supply and the former neutral conductors provide the negative d.c supply.

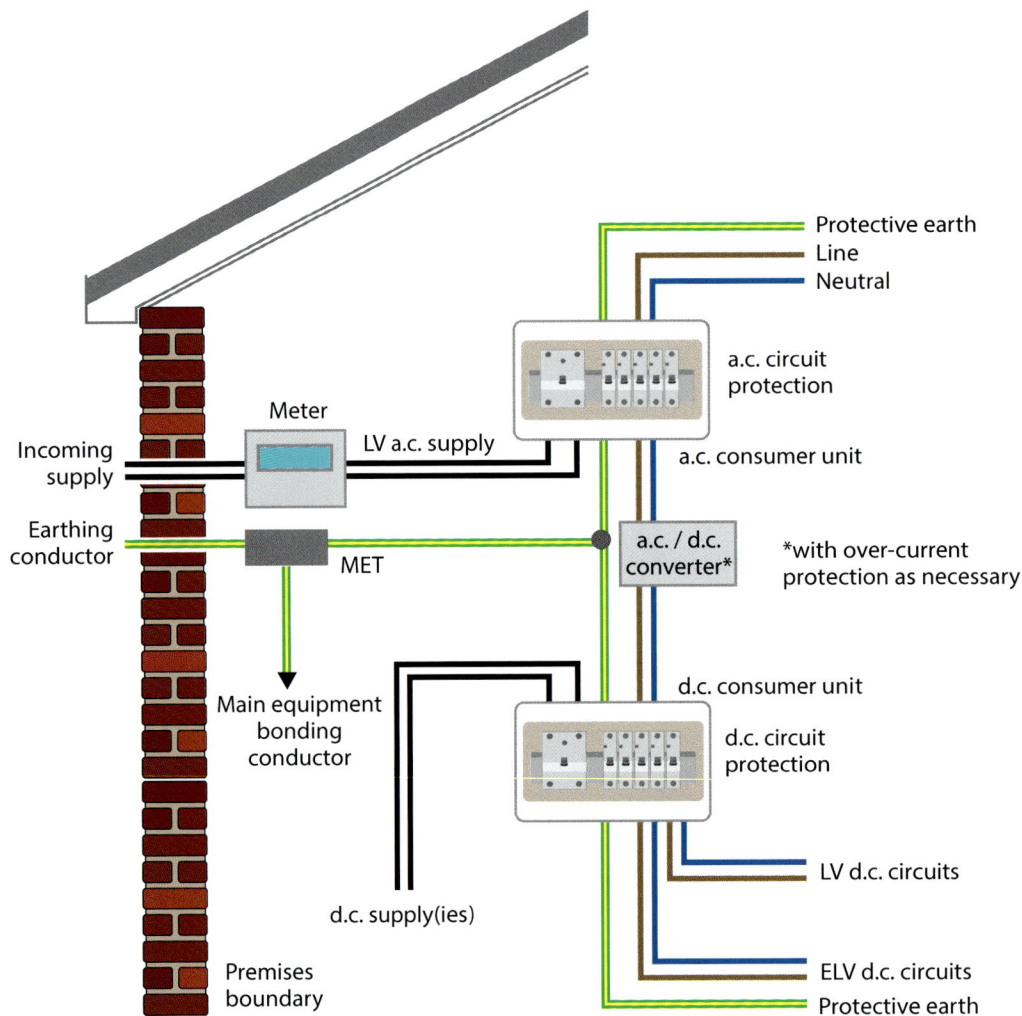

Figure 8: LV and ELV d.c. sub-systems within a.c. building and protected from a.c. and d.c. side

48

© The Institution of Engineering and Technology

7.4.4   Isolation

Where required, isolation should be provided by devices that are appropriately rated for the d.c. circuit that is provided.

NOTE: Where a.c. devices are installed, correct de-rating factors should be applied.

7.4.5   Over-current protection

If a circuit breaker specified for a.c and d.c. is to be used for d.c. protection, its operating characteristics should be assessed to ensure its suitability. The specification and application of circuit breakers within an appropriate d.c. consumer unit is largely similar to the a.c. case and many of these devices are designed for both a.c. and d.c. applications (although not at the same rating).

LV circuit breakers are covered by the following British Standards:

(a)   BS EN 60898-1 covering a.c. circuit breakers for domestic or similar use;
(b)   BS EN 60898-2 covering a.c. and d.c. circuit breakers for domestic or similar use;
(c)   BS EN 60947-2 (industrial or similar use)

NOTE 1: BS EN 60898-2 is of primary interest in this Section, as the standard is used in conjunction BS EN 60898-1 as it specifies the modifications to BS EN 60898-1 for use in d.c. applications. Circuit breakers for use in d.c. applications must therefore be compliant with BS EN 60898-2, which specifies additional requirements for d.c. applications.
NOTE 2: Only Types B and C circuit breakers are specified for d.c. use.

Selection of the appropriate device for d.c. application follows the same process as in the case of a.c. and is based upon:

(d)   the type of circuit breaker (B or C), which depends on the load type (purely resistive, purely inductive or mixed);
(e)   the rated current and rated short circuit capacity.

For earthed systems, the maximum earth-loop impedance (Zs) should be used in accordance with BS 7671.

As for a.c. circuits, it is possible to use circuit breakers in accordance with BS EN 60898-2 in conjunction with a residual current device. Information on the implementation for residual current devices for d.c. distribution is not available.

Some circuit breakers that have been designed specifically for LV d.c. applications feature permanent magnetic elements that establish positive and negative polarities, and the connections of such types of breakers have to be in compliance with the polarities. Incorrect connections of the polarities can damage the breaker.

7.4.6 Testing

Following completion of the conversion tasks, each installation (i.e. the a.c. installation and the d.c. installation) shall be verified in accordance with BS 7671.

NOTE 1: Loop resistance values should be compared with installation design maximum values.

NOTE 2: Insulation resistance values < 2 MΩ may warrant investigation.

7.5   Co-existence issues

7.5.1   Safety

The possibility of electrical faults between multi-level d.c. and a.c. supplies may require further study.

If bipolar d.c. distribution circuits are created, then equipment should have adequate protection against unbalanced and excessive voltage on the designated '+/–' conductors resulting from failure of the '0' designated conductor.

7.5.2   Power quality

The impact of high discharging currents (with peaks up to kA) from storage devices and the smoothing capacitors of a.c./d.c. converters during a fault on adjacent equipment study should be considered at the design stage (d.c. fuses and circuit breakers cannot protect against this).

7.5.3   Earthing

See Section 54 of BS 7671.

7.5.4   Cable segregation and identification

See 7.1.5, 7.4.3 and 7.6.

7.6   Labelling

d.c. conductors shall be coloured, marked or otherwise labelled to identify the function and polarity of the conductors in accordance with BS 7671.

New fixed wiring installations can be specified for the use of cables formerly providing single a.c. circuits, labelled to identify their use for ELV d.c. or LV d.c.

The internal surfaces of switch housings shall be labelled to indicate the presence of d.c. supply and the conductors should be labelled to indicate supply polarity.

NOTE: The general convention is that the former line conductors provide the positive d.c. supply and the former neutral conductors provide the negative d.c. supply.

7.7 Certification protocol

Appropriate certification documentation compliant to local, regional or national regulations should be completed for both retrofit and new installations and would apply to the entire installation i.e. both new and existing wiring.

In addition, for the mains a.c. installation a standard Electrical Installation Certificate should be present or, in the case of retrofit, a current Electrical Installation Condition Report will suffice. In any case the installer should be aware of their responsibilities as outlined above (System Installation and Testing) and declare the extent and limitations of their work on the Electrical Installation Certificate.

(a)   Skilled persons, electrically (as defined in BS 7671), as appropriate to their function will have sound knowledge and experience relevant to the nature of the work undertaken and to the technical standards set down in these Regulations, be fully versed in the inspection and testing procedures contained in these Regulations and employ adequate testing equipment.

(b)   Certificates will indicate the responsibility for design, construction, inspection and testing, whether in relation to new work or further work on an existing installation.

(c)   Schedules of inspection and schedules of test results should be issued with the Electrical Installation Certificate.

(d)   When filling in and signing a form on behalf of a company or other business entity, individuals should state for whom they are acting.

(e)   Additional forms may be required as clarification, if needed by ordinary persons, or in expansion for larger or more complex installations.

(f)   Documentation, including system schematics and specification, shall be provided to the client with a copy located close to the installation and another kept by the installer.

Proprietary d.c. power distribution over conventional 3-phase a.c. power supply cabling

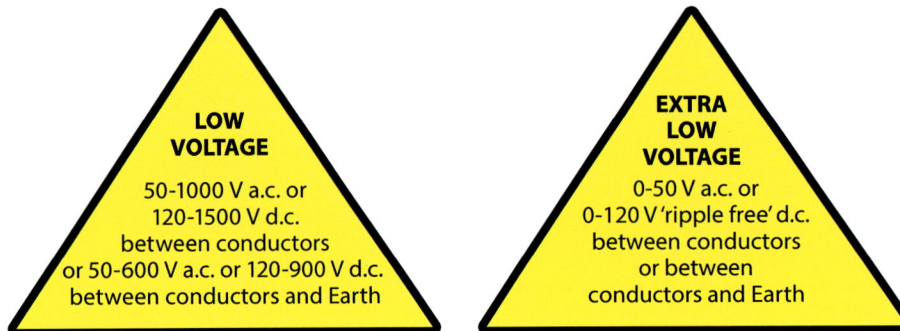

LOW VOLTAGE

50-1000 V a.c. or
120-1500 V d.c.
between conductors
or 50-600 V a.c. or 120-900 V d.c.
between conductors and Earth

EXTRA LOW VOLTAGE

0-50 V a.c. or
0-120 V 'ripple free' d.c.
between conductors
or between
conductors and Earth

8.1   General

8.1.1   Basic principles

3-phase a.c. power distribution circuits containing multiple line and neutral conductors may be converted to the distribution of ELV or LV d.c. provided that each d.c. circuit uses two (unipolar d.c.) or three (bipolar d.c.) conductors and does not share the conductors of another d.c. circuit.

NOTE: Different design of a.c./d.c. converters is required to form and control unipolar or bipolar d.c. systems.

As a result, it should be noted that the network topology of a 3-phase a.c. circuit will be impacted by conversion to d.c. i.e. a four-conductor 3-phase circuit serving three separate single-phase circuits will result in only two unipolar d.c. circuits (as exemplified in Figure 9). As a result great care is required in planning any such conversion.

Earthing arrangements of converted circuits should be assessed for:

(a)   suitability to support d.c distribution;
(b)   compliance with current practice as defined in BS 7671.

Existing earthing conductors should be not used as d.c circuit conductors.

8.1.2   Section structure

The structure of this Section is as follows:

(a)   8.2 describes the testing of the a.c. power distribution cabling before initiating any action to convert it to d.c. power distribution;
(b)   8.3 describes the method of determining the capability of the selected a.c. power distribution cabling to support d.c. power distribution;

(c)   8.4 describes the implementation of the d.c. power distribution over the selected a.c. power distribution cabling including migration of consumer units etc.;

(d)   8.5 discusses the co-existence of d.c. and a.c. circuits;

(e)   8.6 provides information about labelling;

(f)   8.7 provides additional information about certification of the d.c. power distribution installation.

Safe working practices are described in HSG85.

8.1.3   Procedures

See 7.1.3.

8.1.4   Certification

See 7.1.4.

8.1.5   Mixing ELV and LV circuits

See 7.1.5.

8.2   Verification of existing a.c. power supply cabling

See 7.2.

8.3   Capability assessment

See 7.4.

8.4 Implementation

8.4.1 Simple installations and final circuits

8.4.1.1 General

This sub-clause provides diagrams showing conductors within a cable construction. The general principles outlined in Figure 9 to Figure 12 are applicable to the numbers of conductors shown when presented as single or other combinations of conductors.

There is no general convention for designating which of the conductors provide the positive d.c. supply/supplies and which of the former neutral conductors provide the negative d.c. supply/supplies.

8.4.1.2 TN-S and TN-C-S systems

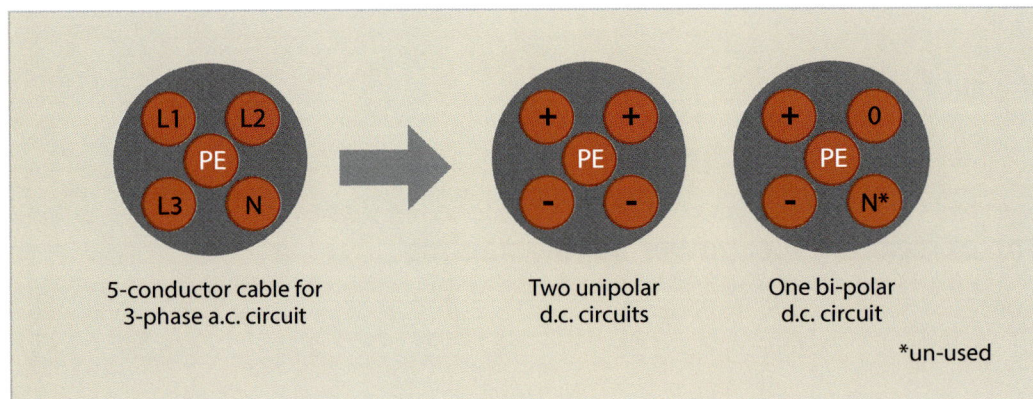

5-conductor cable for
3-phase a.c. circuit

Two unipolar
d.c. circuits

One bi-polar
d.c. circuit

*un-used

Figure 9:
5-conductor cable
for 3-phase a.c.
circuit converted
to d.c. installation

NOTE: The size of a neutral conductor in 3-phase circuits may be not identical to that of the phase conductors. If the neutral conductor is used to form a parallel path for a d.c. unipolar system as shown in Figure 9, then the current carrying capacity will be limited by the cross-sectional area of the neutral, which, in some 3-phase systems, can be smaller than the phase conductor.

NOTE: Unused conductors should be subjected to the safety practices defined in BS 7671.

8.4.1.3 TN-C systems

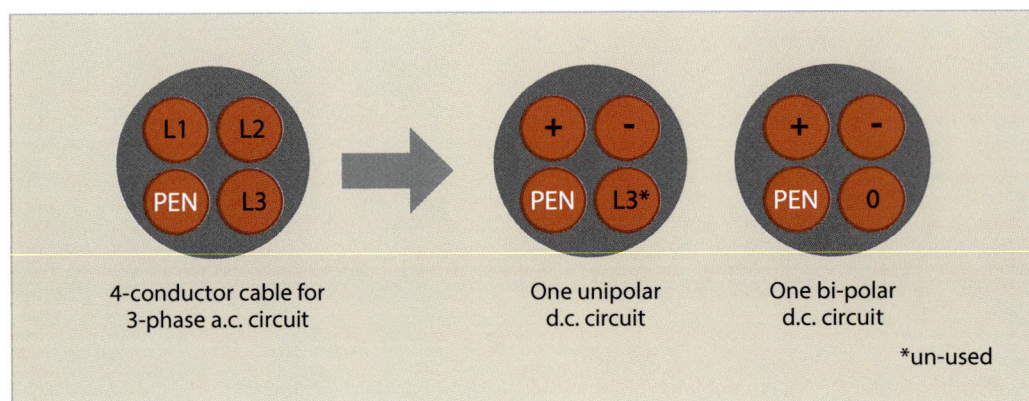

4-conductor cable for
3-phase a.c. circuit

One unipolar
d.c. circuit

One bi-polar
d.c. circuit

*un-used

Figure 10:
4-conductor cable
for 3-phase a.c.
circuit converted
to d.c. installation

NOTE: Unused conductors should be subjected to the safety practices defined in BS 7671.

8.4.1.4 TT systems

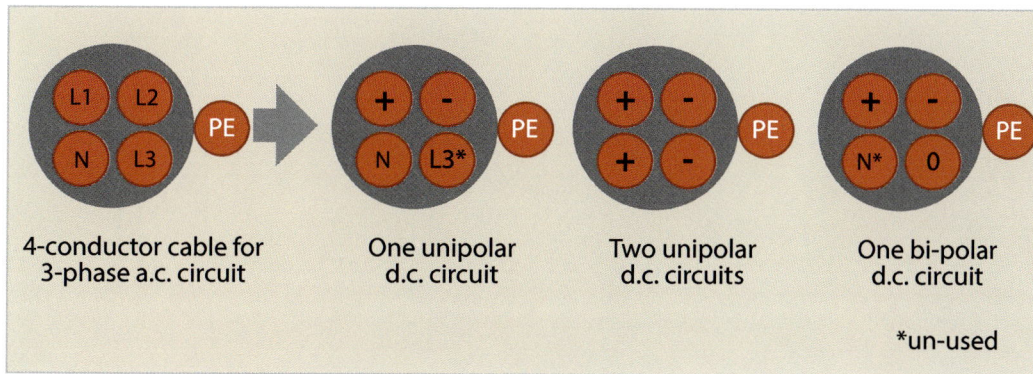

Figure 11: 4-conductor cable for 3-phase a.c. circuit converted to d.c. installation

4-conductor cable for 3-phase a.c. circuit

One unipolar d.c. circuit

Two unipolar d.c. circuits

One bi-polar d.c. circuit

*un-used

NOTE: The size of a neutral conductor in 3-phase circuits may not be identical to that of the phase conductors. If the neutral conductor is used to form a parallel path for a d.c. unipolar system as shown in Figure 11, then the current carrying capacity will be limited by the cross-sectional area of the neutral, which can be in some 3-phase systems smaller than the phase conductor.

NOTE: Unused conductors should be subjected to the safety practices defined in BS 7671.

8.4.1.4 IT systems

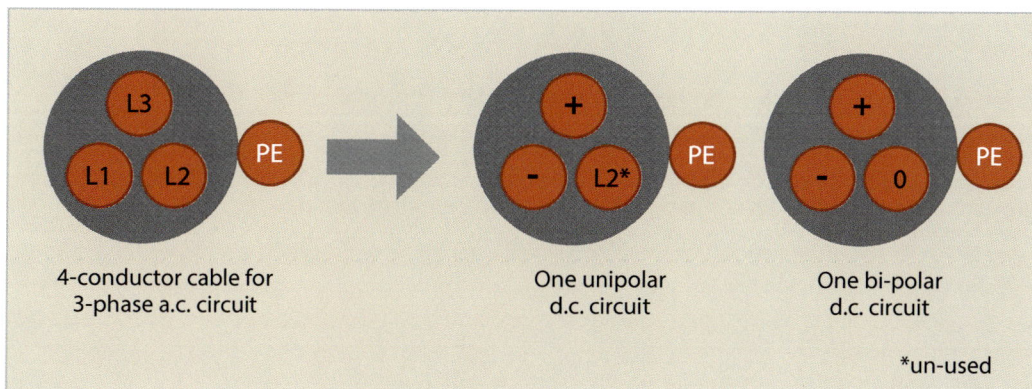

Figure 12: 3-conductor cable for 3-phase a.c. circuit converted to a d.c. installation

4-conductor cable for 3-phase a.c. circuit

One unipolar d.c. circuit

One bi-polar d.c. circuit

*un-used

NOTE: Unused conductors should be subjected to the safety practices defined in BS 7671.

8.4.2 d.c. power source segregation

See 7.4.2.

8.4.3 Migration issues

See 7.4.3.

8.4.4 Isolation

See 7.4.4.

8.4.5 Over-current protection

See 7.4.5.

8.4.6 Testing

See 7.4.6.

© The Institution of Engineering and Technology

8.5   Co-existence issues

8.5.1   Safety

See 7.5.1.

8.5.2   Power quality

See 7.5.2.

8.5.3   Earthing

See 7.5.3.

8.5.4   Cable segregation and identification

See 7.5.4.

8.6   Labelling

d.c. conductors shall be coloured, marked or otherwise labelled to identify the function and polarity of the conductors in accordance with BS 7671.

New fixed wiring installations can be specified for the use of cables formerly providing single a.c. circuits, labelled to identify their use for ELV d.c. or LV d.c.

The internal surfaces of switch housings shall be labelled to indicate the presence of d.c. supply and the conductors should be labelled to indicate supply polarity.

NOTE: There is no general convention for designating which of the conductors provide the positive d.c. supply/supplies and which of the former neutral conductors provide the negative d.c. supply/supplies.

8.7   Certification protocol

See 7.7.

Relevant standards

ANSI/TIA-568 series

BS 7671:2008+A3:2015, *Requirements for Electrical Installations. IET Wiring Regulations*

BS EN 50173 series, *Information technology – Generic cabling systems*

BS EN 50174 series, *Information technology – Cabling installation*

BS EN 50272 series, *Safety requirements for secondary batteries and battery installations*

BS EN 50310 *Application of equipotential bonding and earthing in buildings with information technology equipment*

BS EN 60512-9-3, *Connectors for electronic equipment – Tests and measurements Part 9-3: Endurance tests – Test 9c: Mechanical operation (engaging/separating) with electrical load*

BS EN 60512-99-001, *Connectors for electronic equipment – Tests and measurements – Part 99-001: Test schedule for engaging and separating connectors under electrical load – Test 99a: Connectors used in twisted pair communication cabling with remote power*

BS EN 60898-1, *Electrical accessories – Circuit breakers for overcurrent protection for household and similar installations – Part 1: Circuit-breakers for a.c. operation*

BS EN 60898-2, *Electrical accessories – Circuit-breakers for overcurrent protection for household and similar installations – Part 2: Circuit-breakers for a.c and d.c. operation*

BS EN 60950-1, *Information technology equipment – Safety Part 1: General requirements*

BS PD CLC TR 50174-99-1: *Information technology – Cabling installation: Remote powering*

ETSI EN 300 132-3-1, *Environmental Engineering (EE);Power supply interface at the input to telecommunications and datacom (ICT) equipment; Part 3: Operated by rectified current source, alternating current source or direct current source up to 400 V; Sub-part 1: Direct current source up to 400 V*

ETSI EN 301 605, *Information Environmental Engineering (EE); Earthing and bonding of 400 VDC data and telecom (ICT) equipment*

ISO/IEC 11801 Ed.2.2:2011, *Information technology – Generic cabling systems for customer premises*

IEEE 802.3at:2009: *Standard for Information Technology – Telecommunications and information exchange between systems – Local and metropolitan area networks – Specific Requirements; Part 3: Carrier Sense Multiple Access with Collision Detection (CSMA/CD) Access Method and Physical Layer Specification: Amendment 3: Data Terminal Equipment (DTE)* Power via the Media Dependent Interface *(MDI) Enhancements*

IEEE 802-3bt*

ISO/IEC 29125:2010 *Information technology – Telecommunications cabling requirements for remote powering of terminal equipment*

* in preparation at this time

Bibliography

HM Government, The Building Regulations 2010 Approval Document P: *Electrical safety – Dwellings*

HM Government, Statutory Instrument 1994 No. 1768, The Plugs and sockets etc. (Safety) Regulations 1994

Health and Safety Executive, HSG 85 (Health & Safety Executive Guide), *Electricity at work: Safe working practices*, 2013

IET *Code of Practice for Grid Connected Solar Photovoltaic (PV) Systems**

* in preparation at this time

INDEX

IET

IET Standards

Industry-leading standards

IET Standards works with industry-leading bodies and experts to publish a range of codes of practice and guidance materials for professional engineers, using its expertise to achieve consensus on best practice in both emerging and established technology fields.

See the full range of IET Standards titles at:

www.theiet.org/standards

IET Centre of Excellence

The Institution of Engineering and Technology

IET Centres of Excellence

The IET recognises training providers who consistently achieve high standards of training delivery for electrical installers and contractors on a range of courses at craft and technician levels.

Using an IET Centre of Excellence to meet your training needs provides you with:

- Courses that have a rigorous external QA process to ensure the best quality training
- Courses that underpin the expertise required of the IET Electrical Regulations publications
- Training by competent and professional trainers approved by industry experts at the IET

See the current list of IET Centres of Excellence in your area at:

www.theiet.org/excellence

IET

Electrical **excellence**

Expert publications

The IET is co-publisher of BS 7671 (IET Wiring Regulations), the national standard to which all electrical installations should conform. The IET also publishes a range of expert guidance supporting the Wiring Regulations.

You can view our entire range of titles including...

- BS 7671
- Guides
- Guidance Notes series
- Inspection, Testing and Maintenance titles
- City & Guilds textbooks and exam guides

...and more at:

www.**theiet**.org/electrical

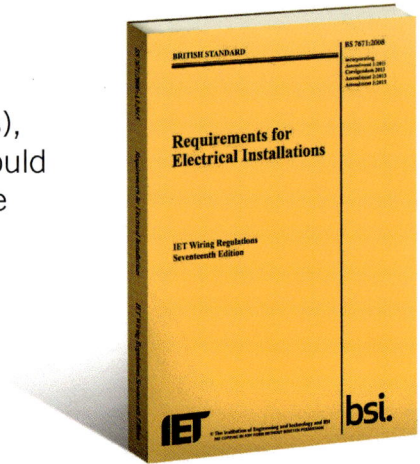

ELECTRICAL STANDARDS ➕

Constantly up-to-date digital subscriptions

Our expert content is also available through a digital subscription to the IET's Electrical Standards Plus platform. A subscription always provides the newest content, giving peace of mind that you are always working to the latest guidance.

It also lets you spread the cost of updating all your books once new versions are released.

Going digital gives you greater flexibility when working with the Wiring Regulations, Guidance Notes and the IET's expert Codes of Practice available for electrical engineers. The intuitive search function, instantly serves results from across all books in your package. You can also access the content on your desktop, laptop or tablet, making it easy to take the content out on site or read on the move.

Find out more about our subscription packages and choose one to suit you at:

www.**theiet**.org/esplus